How Everything Happened

Including Us

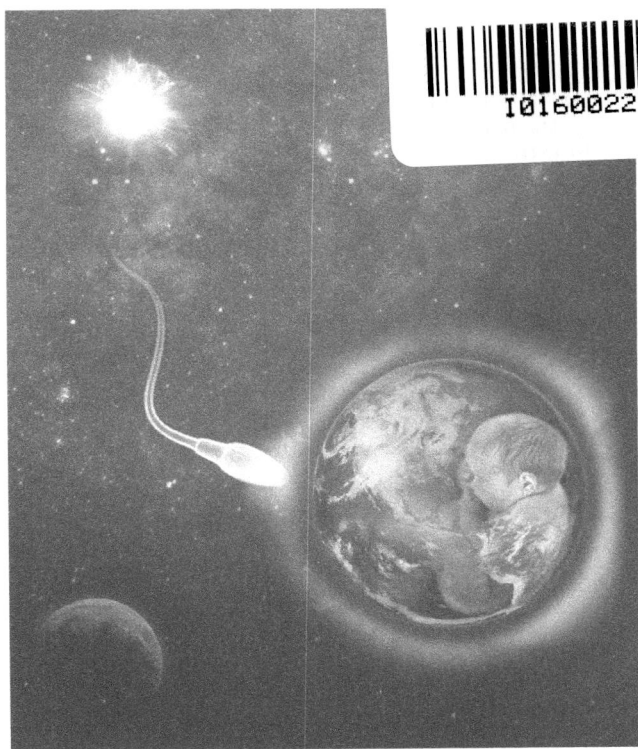

Larry Bell

How Everything Happened: Including Us

Other books by Larry Bell:

Scared Witless: Prophets and Profits of Climate Doom
Climate of Corruption: Politics and Power behind the Global
Warming Hoax
Cosmic Musings: Contemplating Life beyond Self
Reflections on Oceans and Puddles: One Hundred Reasons to be
Enthusiastic, Grateful and Hopeful
Thinking Whole: Rejecting Half-Witted Left & Right Brain
Limitations
Reinventing Ourselves: How Technology is Rapidly and Radically
Transforming Humanity
Cyberwarfare: Targeting America, Our Infrastructure and Our
Future

STAIRWAY PRESS—Apache Junction

Book Cover Art: Tamalee Basu
Cover Design by Chris Benson
www.BensonCreative.com

STAIRWAY⹀PRESS
www.StairwayPress.com
1000 West Apache Trail, Suite 126
Apache Junction, AZ 85120

How Everything Happened: Including Us

Other books by Larry Bell:

Scared Witless: Prophets and Profits of Climate Doom
Climate of Corruption: Politics and Power behind the Global Warming Hoax
Cosmic Musings: Contemplating Life beyond Self
Reflections on Oceans and Puddles: One Hundred Reasons to be Enthusiastic, Grateful and Hopeful
Thinking Whole: Rejecting Half-Witted Left & Right Brain Limitations
Reinventing Ourselves: How Technology is Rapidly and Radically Transforming Humanity
Cyberwarfare: Targeting America, Our Infrastructure and Our Future

STAIRWAY PRESS—Apache Junction

Book Cover Art: Tamalee Basu
Cover Design by Chris Benson
www.BensonCreative.com

STAIRWAY≡PRESS
www.StairwayPress.com
1000 West Apache Trail, Suite 126
Apache Junction, AZ 85120

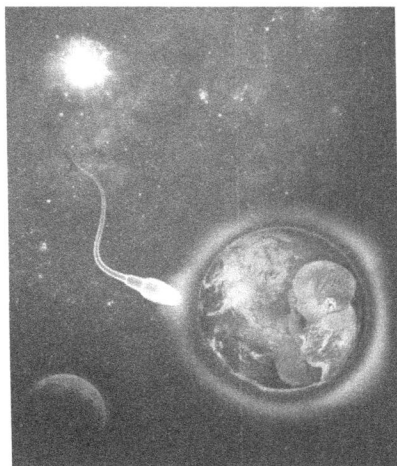

Dedicated with profound appreciation,
amazement and humility to everything
that somehow miraculously transpired to
make this book and its author happen.

About the Author

LARRY BELL IS an endowed professor of space architecture at the University of Houston where he founded the Sasakawa International Center for Space Architecture (SICSA) and the graduate program in space architecture.

Larry's other recent books include: *Thinking Whole: Rejecting Half-Witted Left & Right Brain Limitations* (2018), *Reflections on Oceans and Puddles: One Hundred Reasons to be Enthusiastic, Grateful and Hopeful* (2017), *Cosmic Musings: Contemplating Life Beyond Self* (2016), *Scared Witless: Prophets and Profits of Climate Doom* (2015), and *Climate of Corruption: Politics and Power Behind the Global Warming Hoax* (2011). He is currently working on a new book co-authored with Buzz Aldrin, *Beyond Footprints and Flagpoles.*

Larry is also an entrepreneur who has co-founded several U.S. and international commercial space companies. One—established with NASA's Chief Engineer and two other partners—grew through mergers and acquisitions to employ more than 8,000 professionals, went on the New York Stock Exchange, and was purchased by General Dynamics.

Professor Bell's many national and international honors include two awards granted by Russia's Institute of Aeronautics and Astronautics: The Konstantin Tsiolkovsky Medal and the Yuri Gagarin Diploma.

Larry's name was placed in large letters on the Russian rocket that launched the first crew to the International Space Station.

Introduction

UNLIKE OTHER OF my books and numerous articles which cover topics I confidently claim to know a great deal about in advance—space technology, energy systems and climate science, for example—this project was different. Broadly stated, my overarching purpose was, first and foremost, a personal learning, contemplation and sharing adventure.

As promised in the title, this book addresses a mind-numbingly expansive scope and time scale of historical events. Some occurred over billions of years, others in cosmic blinks; some were unfathomably large, others imperceptivity small; some had catastrophic impacts upon certain regions and species, benefiting others more fortunate; and some were willfully tragic, while others spiritually inspiring.

In all cases, these events are now history. The results are self-evident. We can't change them.

The best we can do is attempt to draw constructive lessons regarding what and how things happened in the past to influence tomorrow's history.

Had things turned out quite differently, I wouldn't be here to write this book in my time-space, nor would you be reading this sentence now in yours.

Just as I consciously attempt to avoid unnecessary moralistic editorializing regarding the irreversible past, neither is it possible to refrain from discussing horrific travesties perpetrated on pretexts of imperial and divine authorities. Concurrently, I will further clarify my interpretations of the entitled terms "how" everything happened, and who the "us" it happened to are.

Explicitly, the "how" issue exclusively describes only causal connections that account for the occurrence of any one particular series of events. Very importantly, I do not presume to conjecture larger religious or philosophical explanations regarding "why" they occurred.

The "us" reference in this book title attempts to focus on ALL of us irrespective of where we lie on the globe—not just North Americans. However, having said this, readers should be forewarned that my inevitable personal bias is bound to influence what is most important to write about, when and where it first happened and who was most important in causing it.

I do promise that this will be a remarkably wondrous story—certainly not because I chronicled it—but rather because against all rational odds, it actually happened and continues to eternally unfold beyond measurements and complexities of human comprehension.

The story begins a very long time ago, 13.7 billion years ago, about 5.3 billion before our planet was born 4.5 billion years ago.

It then took about another 4 billion years for Earth to become teeming with simple, single-celled organisms which eventually evolved into you and me. This involved a very long process as well.

It took 800 million years for those animated self-replicating microdots to transform into such complex living marvels as ferns, flowers, cephalopods, dinosaurs, reptiles and a branch of vertebrates we now call mammals.

The world's ecosystems experienced some big setbacks along the evolutionary highway. Five separate mass extinctions attributed to major climate changes, catastrophic asteroid impacts and massive volcanic eruptions variously killed between three-quarters and 96 percent of all life species.

A sub-branch of some of those very lucky surviving mammals eventually grew brain frontal lobes capable of self-awareness and tribal social behaviors. About 300,000 years ago, a single branch of these "primates" underwent a genetic mutation that allowed speech.

Within only the last ten thousand years some of those Homo sapiens ancestors of ours invented agriculture; battled and domesticated larger animals for food and clothing; competitively warred against each other and Neanderthal hunter-gatherers; established settlements, cities and empires, built great pyramids and cathedrals; formulated complex cultures and laws; developed advanced scientific methods and philosophies; and composed inspirational literature, music and sonnets.

Some inventive and adventuresome Sapiens contemplated the architecture and workings of a celestial universe and applied that knowledge to guide voyages of discovery, trade, conquest and migration to extend domains and dominions.

Others, within little more than the last century, have harnessed the power of lightning and atoms; mastered flight; traveled many times faster than the speed of sound; transmitted information from everywhere to everywhere else via orbital satellites; walked on the Moon and conceived artificial brains that can already outsmart their human creators.

And, for better or worse, we're now really only getting started.

Albert Einstein

Werner Heisenberg

Neils Bohr

Max Planck

Stephen Hawking

Albert Einstein:
General Theory of Relativity

Werner Heisenberg:
Heisenberg's uncertainty principle

Max Planck:
Originator of Quantum Theory

Neils Bohr:
Atomic structure, Quantum Theory

Stephen Hawking:
Gravitational Singularity Theorems

STELLAR CONTRIBUTORS IN QUANTUM PHYSICS

Larry Bell

How Everything Happened, Including Us

Larry Bell

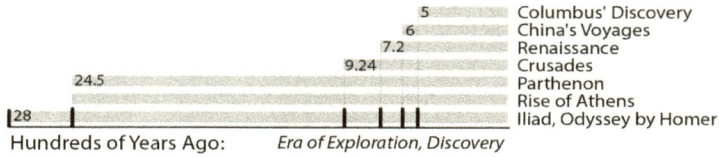

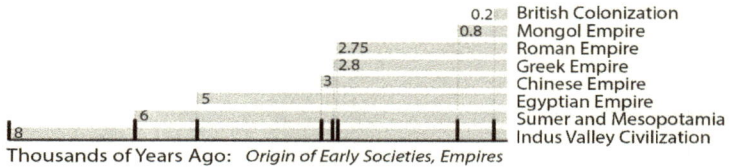

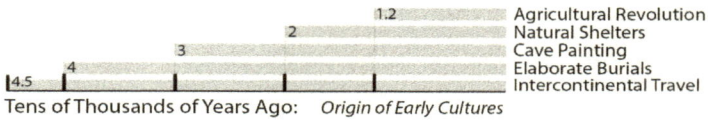

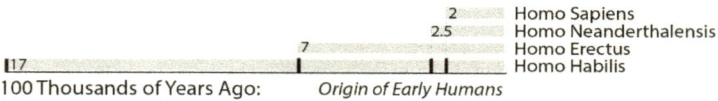

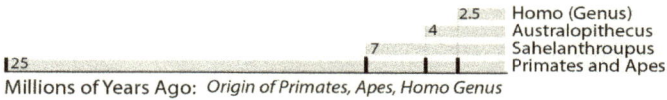

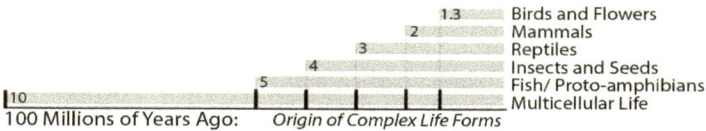

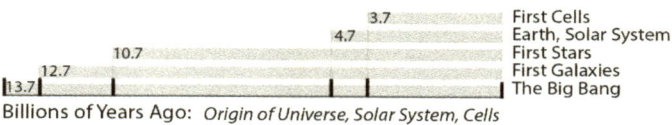

5	Columbus' Discovery
6	China's Voyages
7.2	Renaissance
9.24	Crusades
24.5	Parthenon
	Rise of Athens
28	Iliad, Odyssey by Homer

Hundreds of Years Ago: *Era of Exploration, Discovery*

0.2	British Colonization
0.8	Mongol Empire
2.75	Roman Empire
2.8	Greek Empire
3	Chinese Empire
5	Egyptian Empire
6	Sumer and Mesopotamia
8	Indus Valley Civilization

Thousands of Years Ago: *Origin of Early Societies, Empires*

1.2	Agricultural Revolution
2	Natural Shelters
3	Cave Painting
4	Elaborate Burials
4.5	Intercontinental Travel

Tens of Thousands of Years Ago: *Origin of Early Cultures*

2	Homo Sapiens
2.5	Homo Neanderthalensis
7	Homo Erectus
17	Homo Habilis

100 Thousands of Years Ago: *Origin of Early Humans*

2.5	Homo (Genus)
4	Australopithecus
7	Sahelanthropus
25	Primates and Apes

Millions of Years Ago: *Origin of Primates, Apes, Homo Genus*

1.3	Birds and Flowers
2	Mammals
3	Reptiles
4	Insects and Seeds
5	Fish/ Proto-amphibians
10	Multicellular Life

100 Millions of Years Ago: *Origin of Complex Life Forms*

3.7	First Cells
4.7	Earth, Solar System
10.7	First Stars
12.7	First Galaxies
13.7	The Big Bang

Billions of Years Ago: *Origin of Universe, Solar System, Cells*

KEY EVENT TIMELINES

4

How Everything Happened, Including Us

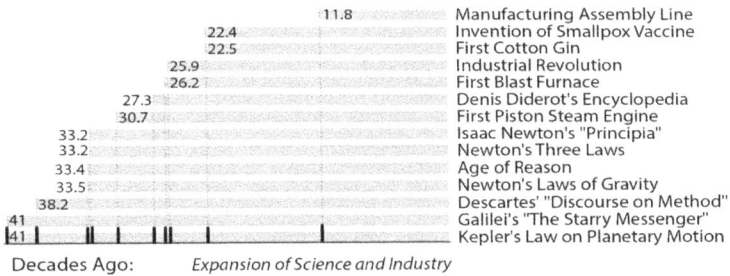

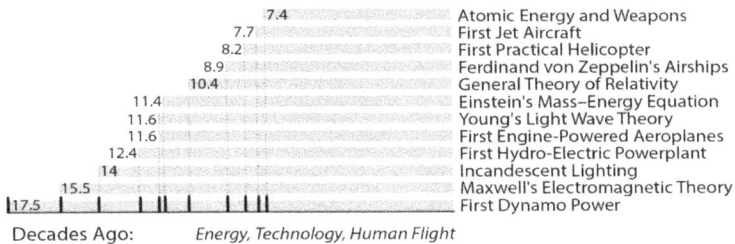

DECADES AGO

| Expansion of Science and Industry | Energy, Technology, Human Flight | World Wars and Campaigns | The Space Age | The Information Revolution |

	11.8	Manufacturing Assembly Line
22.4		Invention of Smallpox Vaccine
22.5		First Cotton Gin
25.9		Industrial Revolution
26.2		First Blast Furnace
27.3		Denis Diderot's Encyclopedia
30.7		First Piston Steam Engine
33.2		Isaac Newton's "Principia"
33.2		Newton's Three Laws
33.4		Age of Reason
33.5		Newton's Laws of Gravity
38.2		Descartes' "Discourse on Method"
41		Galilei's "The Starry Messenger"
41		Kepler's Law on Planetary Motion

Decades Ago: *Expansion of Science and Industry*

	7.4	Atomic Energy and Weapons
7.7		First Jet Aircraft
8.2		First Practical Helicopter
8.9		Ferdinand von Zeppelin's Airships
10.4		General Theory of Relativity
11.4		Einstein's Mass–Energy Equation
11.6		Young's Light Wave Theory
11.6		First Engine-Powered Aeroplanes
12.4		First Hydro-Electric Powerplant
14		Incandescent Lighting
15.5		Maxwell's Electromagnetic Theory
17.5		First Dynamo Power

Decades Ago: *Energy, Technology, Human Flight*

KEY EVENT TIMELINES

5

Decades Ago	Event
6.4	Warsaw Pact
6.9	First Hydrogen Bomb
7	Formation of NATO
7.4	Yalta Conference
7.7	Battle of Atlantic
7.7	Battle of Guadalcanal
7.8	Stalin's attack on USSR
8	Second World War
8.3	Spanish Civil War
8.3	Great Stalin Purge
8.6	Adolf Hitler Era
9	Great Depression
10.1	Second Battle of Marne
10.2	Russian Revolution
10.3	Western Front Campaigns
10.5	First World War

Decades Ago: *World Wars and Campaigns*

Decades Ago	Event
0	First Black Hole Image
0.4	First Reusable Rocket- Falcon 9
0.5	First Comet Landing- Rosetta
0.6	First Object Leaves Solar System
2.1	International Space Station
2.4	First Probe on Jupiter
3.3	First Space Station- Mir
3.8	First Reusable Space Shuttle
4.2	Voyager 1 & 2
4.4	First Probe on Venus- Venera 7
5	First Humans on Moon
5.3	First Lunar Probe- Luna 9
5.8	First Human in Space
6.2	First Orbital Object- Sputnik
9.3	

Decades Ago: *The Space Age*

Decades Ago	Event
0.6	Google Quantum AI Laboratory
0.8	Computer Wins Jeopardy Game
0.8	Commercial Quantum Computer
2.2	Computer Wins Chess Game
3.6	Internet Invented
4.8	First Personal Computers
4.9	Cloud Computing
5	Commercial Memory Chip
6.3	"Thinking Machine" Concept

Decades Ago: *The Information Revolution*

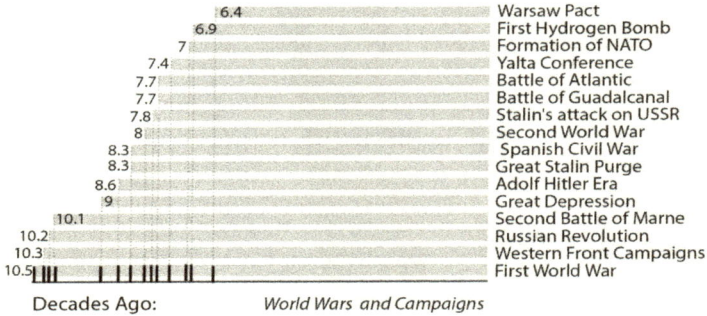

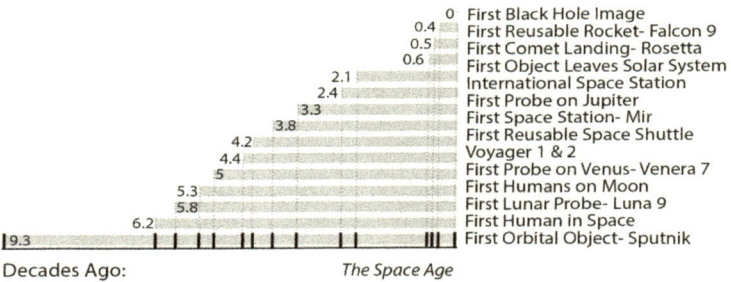

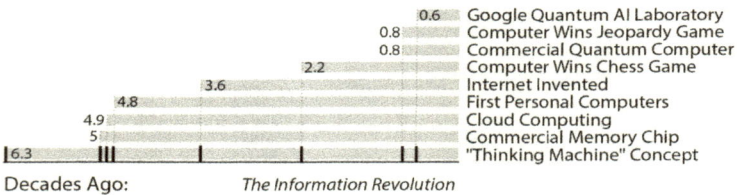

KEY EVENT TIMELINES

6

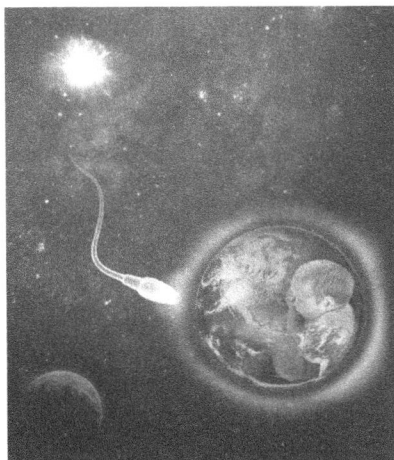

Part One: Our Living Universe(s)

WE ARE TRULY creatures of the natural Universe. The elements that constitute every part of us and our beautiful blue planet passed through other fiery stars, now gone, long before our mother Sun was born. Each of those elements, in turn, are comprised of atoms which are like tiny solar systems containing tremendous energy, each with a proton and neutron nucleus orbited by electrons that together direct the complex chemistry of life.

Astronomer Sir Martin Rees suggested that those atoms may last 10^{35} years. This would clearly indicate that the atomic and subatomic building blocks of our bodies were once subatomic particles of the primordial cosmos and interstellar clouds of our Milky Way galaxy which have only recently passed through myriads of rock formations and organisms on Earth.

And as *Out of Chaos, Evolution from the Big Bang to Human Intellect* author Wayne M. Bundy suggests:

> *Recycling through single-celled life, ancient*

hominids, such notables as Alexander the Great, saints and thieves, and many more, our atoms have seen it all.[1]

The known Universe that spawned us has existed about 13.7 billion years, about 9 billion years longer than our home planet, and is constantly changing, like a garden where new plants bud as others wither on a cosmic time schedule.

Wayne Bundy urges us to recognize that "Our fundamental commonality with all existence arises from these processes...we are the stuff of stars."

Yet from our vantage point on a spiral arm of our Milky Way galaxy, it is difficult to grasp the reality that those distant stars, and the planets and ghostly clouds that surround them, are part of our personal world.

Humankind has a history of resisting any observations that placed us outside the center of the Universe. And in reality, the Universe probably has no center, except maybe a theoretical point where a Big Bang first set everything in motion. Yet there is really nothing to be upset about. The real estate we occupy has a wonderful location suited perfectly to our lives and a spectacular view.

Many people tend to think that this cosmic perspective diminishes human significance. Perhaps you have seen graphic illustrations depicting the Milky Way, possibly printed on a tee shirt, with an arrow pointing to our solar system neighborhood and a note stating "you are here." Some will take this as evidence that we are indeed small-timers living in the celestial boondocks. Others, however, may interpret this very differently, recognizing that we are all an integral part of something unfathomably majestic and empowering.

While it is true that our Earth community and human bodies are small relative to the Universe, we should remember that everything is small relative to the Universe. After all, does size really have anything at all to do with importance? Are boulders more important than hamsters? Is Saturn more important than the Earth? It seems to me that either everything is important, or nothing is. And that is purely a matter of personal decision.

The idea of a changing Universe can also be disquieting for some. If even planets, stars and galaxies are constantly being born and changing, ultimately only to die, then where does our human destiny lie? Upon what permanent ground can we build our spiritual refuge?

One answer is that change is the essential nature of life and human spirit. Everything that we have the good judgment to enjoy is dynamic, revealing new dimensions of possibility with each transformation and discovery. The exciting news, if we accept it, is that nature is eternal, or as close to that as we dare imagine. And as manifestations of that wonderful condition, we are too.

In any event, science does not explain, and likely cannot explain the greatest spiritual marvel of all. Stephen Hawking emphasized that while science can eventually solve the mystery of how the Universe began, it cannot answer the most basic question: why the Universe bothers to exist?

Our local Universe—not necessarily the only one—is unimaginably vast and intellectually astounding. Once thought to be the entire universe, it encompasses some 100 billion galaxies encompassing billions of light years in every direction, each containing 100 billion stars on average. The distances between these galaxies is enormous, typically about 400,000 light years. Most have been speeding away from each other through space expansion at rates directly proportional to their distances apart since the time of the Big Bang. Our Milky Way galaxy and the Andromeda galaxy are exceptions, which speed toward each other at a tremendous velocity.[2]

Our Milky Way galaxy is a brilliant archipelago of more

than 100 billion stars and vast molecular clouds, among an archipelago of galaxies, and perhaps among an endless archipelago of sub-universes. Most of those other stars are larger than our yellow dwarf Sun. Altogether, this spiral wheel is viewed as a flat disc about 100,000 light years in diameter.

If we were somehow able to travel to a new star in our galaxy every hour of the day and night, it would take about six million years, much longer than humans have existed, to visit only about half of them.

Visiting all the planets in our galaxy would take considerably longer, with estimated numbers ranging from 1 billion to 30 billion. That would be far easier, however, than trying to visit all planets in our known Universe where the number would be closer to a billion-billion.

So far as our being alone out here, is our Earth likely to be the only life-sustaining planet?

Let's imagine, for example, that some form of life exists on only one in every billion planets—perhaps about one billion total. Evolutionary biologist Richard Dawkins considers this to be highly unlikely, with life considerably more abundant.[3]

And what about the likelihood of the existence of other intelligent life? Or as the late Stephen Hawking quipped:

> *I believe alien life is quite common in the Universe, although intelligent life is less so. Some say it has yet to appear on planet Earth.*

So first, let's assume that by intelligent life we're referring to creatures we might be able to engage in meaningful communications if not actually enjoy socializing with; particularly if they do not pose a threat to self-esteem and wellbeing. Well, according to a famous Drake Equation, the probable number of such communicating civilizations could be quite enormous.

Formulated by Frank Drake in 1961, the Drake equation

speculates probabilities regarding the number of planetary civilizations by factoring in such considerations as: the rate of star formation; the fraction of those stars with habitable planets; the fraction of those planets which might develop life; the fraction of intelligent life; the further fraction of those capable of developing detectable technology; and finally, the length of time such civilizations might be detectable. The fundamental problem here is that sound statistical estimates are rendered impossible since our own circumstances offer the only known (and highly biased) example.

Those estimates vary widely. In 1966, Carl Sagan optimistically suggested that there might be as many as one million communicating civilizations in the Milky Way alone, although he later decided that number might be far smaller. Mathematical physicists Frank Tipler and John D. Barrow—far more pessimistically—put their average-per-galaxy guess at less than one.

In any case, more than 50 years of SETI radio telescope searches haven't turned up any signs of intelligent alien presence...no evidence that our galaxy is teeming with powerful transmitters other than our own continuously broadcasting and receiving near the 21cm hydrogen frequency as we hoped to find.

Frank Drake described his own equation as just a way of organizing our ignorance on the subject. Still, as a general roadmap of what we need to learn in order to understand this existential question, it has formed the backbone of astrobiology as a science.

According to a Fermi-Hart paradox, it looks like the odds of finding intelligent life on other Earth-like planets aren't great due to the long time required for such beings to evolve, coupled with a relatively short remaining life span of our planet when the Sun brightens as predicted in another billion years or so. Physicists Enrico Fermi and Michael Hart argued that given our Sun is a typical and relatively young star, while billions of stars in the galaxy are billions of years older, we should have seen some evidence of advanced alien civilizations by now.

Hence Fermi's Question: Where is Everybody?

Are they perhaps concealed in other parallel universes and additional unseen dimensions? Some renowned theoretical physicists have attached real credence to such possibilities. Mind-numbingly bizarre quantum theories which apply at subatomic realms indicate that the Universe is far stranger than previous conclusions based exclusively on conventional Newtonian physics suggest.

Weird Alternative Worlds of Quantum Mechanics

Plato's allegory *Phaedo* likens our understanding of the objects and phenomena by which we perceive the world to the experience of emerging from a dark cave into sunlight where only vague shadows of what lies beyond that prison are cast dimly upon the wall. Those shadow forms of perception are both non-physical and non-mental, existing nowhere in time, space, mind or matter.

Quantum mechanics, along with Einstein's Special and General Theories of Relativity, beckon us to move our minds beyond caves of previously bounded references.

Although Newtonian physics works wonderfully well to describe and predict events in our "everyday world," it cannot account for phenomena in the subatomic realm which appear to be governed by very different rules.

Quantum theory explains infinitesimally small worlds, while relativity explains the world of large objects, as billiard balls and planets. Whereas general relativity considers the force of gravity, quantum mechanics limits consideration to the three other known forces: the strong force, the weak force and the electromagnetic force. Unification of relativity and quantum mechanics would require uniting these four forces of nature with more than four space-time dimensions.

How small is subatomic? First, we cannot actually "see" an atom. All scientists can do is speculate about what is there based upon certain observations regarding how atoms tend to behave...which, in turn, depends upon specific means and methods by which they are observed.

Some noted scientists, Stephen Hawking included, have argued that quantum theory supports a simultaneous many-worlds state of reality. One interpretation of this possibility is illustrated by "Schrödinger's Cat," a thought experiment proposed by Viennese physicist Erwin Schrödinger.

Imagine that a cat is insidiously placed inside a sealed box with a device which can release poisonous gas to kill it. There is a 50 percent chance that the random decay of a particular atom in the killing device will trigger the gas to be released within a specified period of time before the box is to be opened. As we cannot witness what happens from outside the box, we will not know if the poor cat is alive or dead until we open it.

We logically assume that there is either good news or bad for the cat even before we open the box...both conditions cannot possibly be true. Strangely, a quantum theory interpretation suggests otherwise, whereby the cat is in a kind of limbo...both alive *and* dead at the same time.

At the instant someone looks into the box, that observer's wave function splits the world into two branches, each with a different edition of the cat. One reality branch has the cat being alive, and the other has the cat exhausting at least one of its nine lives. It follows then, that just as there are different simultaneous editions of the cat in different worlds, there are uncountable editions of us...all of which are equally real.

There are many other interpretations of quantum mechanics...all of which seem unbelievably strange.

As Werner Heisenberg, one of the key quantum mechanics founders, wrote:

Larry Bell

I remember discussions with Bohr [in 1927] which went through many hours till very late at night and ended almost in despair; and when at the end of the discussion I went alone for a walk in the neighborhood park I repeated to myself again and again the question: Can nature possibly be as absurd as it seemed to us in those experiments?[4]

While Einstein made major contributions to the development of quantum mechanics theory, he spent much of his career arguing against it. He nevertheless acknowledged its advantages in explaining subatomic phenomena. Most importantly, it worked.[5]

Perhaps the most astonishing evidence of quantum mechanics weirdness relates to experiments where a single photon of light is projected from a precise position towards two slits on shielding wall in front of a photographic plate. The photographic plate records which slit the photon passes through. Whichever mark was randomly checked first by any observer invariably correlated with the recorded placement of the hit, whereas the other slit option was always blank.

The astounding question is how does light know which slit to pass through at the precise moment it is being observed? There is presently still no way to predict this.

American mathematical physicist and quantum mechanics theorist Henry Stapp wrote:

The central mystery of quantum theory is, how does information get around so quick? How does the particle know there are two slits? How does the information about what is happening elsewhere else get collected to determine what is likely to happen here?[6]

14

Heisenberg is credited with validating a fundamental uncertainty principle. Unlike in the visible world, where Newtonian physics works well, in the subatomic realm it is impossible to know both the exact position and momentum of a particle at the same time. The more we know about one condition, the less we know about the other. We are then left to choose which of these two properties we wish to determine.

Not knowing exactly where a particle is and where it is headed makes it impossible to calculate what will happen next. The only option is to assign a certain probability that a particular group of such particles will likely behave in a certain way...a tendency for something to happen.

As Heisenberg described this uncertain tendency:

> *It was a quantitative version of the old concept of 'potential' in Aristotelian philosophy. It introduced something standing in the middle between the idea of an event and the actual event, a strange kind of physical reality just in the middle between possibility and reality.*[7]

Whereas Newton's laws depict events which are quite easy to understand and picture, probabilities of phenomena addressed by quantum mechanics defy conceptualization, and are impossible to visualize. They also defy logic as we understand it.

Heisenberg wrote:

> *The mathematically-formulated laws of quantum theory show clearly that our ordinary intuitive concepts cannot be unambiguously applied to the smallest particles. All the words or concepts we use to describe ordinary physical objects, such as position, velocity, color, size, and so on, become indefinite and problematic if we try to*

use them of elementary particles.[8]

Traditional science works to construct theories such that for every evidence of absolute truth, there is a corresponding element in the theories. This is not the case with quantum theory.

Newton, for example, applied his investigations to calculate movements of the Moon and planets around the Sun using his own mathematics. His findings matched observations of astronomers, establishing rational celestial mechanics which viewed the Universe as a "Great Machine." Without Newtonian physics, human space programs would never have occurred.

Nevertheless, even Newton could not explain how gravitational influences on static and moving bodies of that Great Machine really worked. As he wrote in his famous *Philosophiae Naturalis Principia Mathematica*:

> *I have not been able to discover the cause of these properties of gravity from phenomena, and I frame no hypotheses...it is good enough that gravity does really exist, and act accordingly to the laws which we have explained, and abundantly serves to account for all the motions of the celestial bodies...*[9]

Newton recognized that the very idea of gravitational forces reaching across space seemed to be foolish. He wrote to fellow scholar Richard Bently:

> *...that one body may act upon another at a distance through a vacuum without the mediation of anything else, by and through which their action and force may be conveyed from one to another, is to me so great an*

absurdity that, I believe, no man who has in philosophic matters a competent faculty of thinking could ever fall into it.[10]

Nevertheless, as Einstein pointed out, quantum mechanics does not replace Newtonian physics:

> *…creating a new theory is not like destroying an old barn and erecting a skyscraper in its place. It is rather like climbing a mountain, gaining new and wider views, discovering unexpected connections between our starting point and its rich environment. But the point from which we started out still exists and can be seen, although it appears smaller and forms a tiny part of our broad view gained by the mastery of the obstacles on our adventurous way up.*[11]

Quantum mechanics has changed the way scientists previously visualized the atom as being much like a tiny model of our Solar System with a Sun-like nucleus in the center. The subatomic nucleus was envisioned to contain nearly all of the atom's mass in the form of positively-charged particles (protons)…along with particles about the same size as the protons but without a charge (neutrons). Hydrogen, a lone exception, is the only atom with no neutrons in its nucleus.

Orbiting the nucleus, like planets orbit the Sun, are electrons containing almost no mass (compared with the nucleus), each having one negative charge. The number of electrons and protons is always the same so that the atom, as a whole, has no charge.

The new quantum theory model presents a vision which is quite different both in reference to substance and scale.

As Henry Stapp explained to the Atomic Energy

Commission:

> ...an elementary particle is not a structure built out of independently existing unanalyzable entities, but rather a web of relationships between elements whose meanings arise wholly from their relationships to the whole...it is, in essence, a set of relationships that reach outward to other things.[12]

Size comparisons of subatomic particles with our Solar System are also dramatically different, whereby distances between an atomic nucleus and its electrons are far greater. According to Ernest Rutherford, who created a model of the atom in 1911, the space occupied by an atom compared with its orbiting electrons are "like a few flies in a cathedral."

Even that model is now obsolete. As described by Gary Zukav, the difference between the atomic level and subatomic level is as great as the difference between the atomic level and the entire planet.

Zukav writes:

> It would be impossible to see the nucleus of an atom the size of a grape. In fact, it would be impossible to see the nucleus of an atom the size of a room. To see a nucleus the size of an atom, the atom would have to be as high as a fourteen-story building![13]

Subatomic particles are not actually real particles made of stuff, yet for convenience are statistically measured as quanta in terms of energy units in the same way as particles. These quanta unceasingly change measurable appearances from energy to mass and back, although within a common identity.

In 1900, Max Planck was the first physicist to calculate the sizes of "energy packets" (quanta) in various waves of light frequency (color) using his mathematical invention famously known as Planck's constant. All of those packets of color, red for example, have the same size.

It should be noted that those waves of light can also be measured as energy particles which do not contain any physical stuff. Depending upon which equipment we select to observe it, some experiments show that light is wave-like, while others show that it is a particle-like phenomenon.

This wave-particle duality paradox presents a perplexing dilemma for scientists who like tidy and definitive answers. Thomas Young's 1903 experiments showed that light must be wave-like, while Einstein proved it is particle-like.

Einstein's theory proposed that light is comprised of tiny particles (photons) analogous to a stream of bullets, whereby energy itself is quantized. He termed this a photoelectric effect.

As Planck described Einstein's theory:

> ...the photons (the 'drops' of energy) do not grow smaller as the energy of the ray grows less; what happens is that their magnitude remains unchanged and they follow each other at greater intervals.[14]

Einstein was not able to dispute the contradiction between light as a wave versus light as quanta, but simply took the contradiction as something which would probably be understood later. Nevertheless, while he is far more famous for two revolutionary theories of relativity, both were based upon his discoveries regarding the quantum nature of light which earned him a Nobel Prize.

In 1913, Danish physicist Niels Bohr came up with an explanation that also earned him a Nobel Prize. Bohr speculated that electrons revolve around the nucleus of an atom at specific distances and in precise orbits, or shells. When an atom such as hydrogen is excited with heat or white light, it causes an electron to jump into one of the outer shells. How far it jumps depends upon how much energy it receives. The electron eventually returns back to shell number one and emits energy in the form of light.

The first of Einstein's breakthrough discoveries, his Special Theory of Relativity, affirms that as with quantum mechanics, appearances of events are relative and dependent upon the observers.

This theory fundamentally tells us three things. First, a moving object measures shorter in the direction of motion as its velocity increases until it reaches the speed of light and disappears. Second, the mass of a moving object measures greater as its velocity increases, until it becomes infinite at the speed of light. Third, moving clocks run more slowly as their velocities increase, until they reach the speed of light.

Here, space and time are not two separate things, but together form space-time where energy and mass are actually different forms of the same thing. How these mass versus energy determinations are measured is influenced by how fast the object and observer are moving relative to one another.

To a very high-velocity space traveler, her wristwatch (if it is a good one), appears to keep perfect time…ticking off sixty seconds each minute. Yet as she accelerates faster and faster, and if her watch also simultaneously communicated its recorded time back to Earth, the reception intervals between seconds, minutes, hours, days and years would grow longer. Ultimately, centuries might pass on Earth during but a few years of her high-velocity experience.

Einstein's remarkable theory is based upon the similarly astounding constancy of the 186,000 miles per second speed that light travels irrespective of any and all observers. Whether measured as a wave or a particle, or whether the velocity of a photon is measured while moving away from or towards an observer, that constant never varies.

Imagine, in contrast, that someone walking at a speed of two miles per hour onboard a train traveling 60 miles per hour is clocked through a window by an observer as it passes a boarding station. If that person is walking in the same direction the train is moving, her measured velocity would be 62 miles per hour...58 if she were walking toward the caboose.

Now compare this with clocking the velocity of a photon fired from the front of that train detected at a stationary target many miles ahead. Logic would predict that by adding the velocity of the train to that of the photon, accurate measurements would record a slightly higher net photon speed. That simply doesn't happen...not even for a photon fired by a rocket ship traveling at 100,000 miles per second as clocked at a recording station on another planet.

Briefly summarized, using light velocity as a measurement constant, Einstein's Special Theory of Relativity demonstrates that two events which happen at the same time in one frame of reference may occur at different times when seen from another frame of reference.

Einstein's realization that moving clocks change their rhythm led Einstein to the conclusion that *now, sooner, later* and *simultaneous* are relative terms which depend upon the state of motion of the observer. This seems counterintuitive, particularly since the time differences are much too small for those of us who live low-velocity Earth-bound existences.

Meanwhile, classical Newtonian laws work just fine until we reach velocities approaching the speed of light. That one-dimensional view of time is entirely separate from our three-dimensional perception of space through which we can locate all objects, ourselves included, in a statically measurable coordinate matrix.[15]

Larry Bell

Einstein's space-time continuum is very different, wherein time is fully integrated as a fourth dimension of reality. Although we cannot actually see that dimension, it serves to define relationships expressible through Pythagorean-like diagrams and mathematical constructions which can be experimentally verified. Those diagrams have not only extended our sense of reality beyond our sensory experiences, but beyond our traditional perceptions of what constitutes common sense as well.

As Gary Zukav observes: "The General Theory of Relativity shows us that our minds follow different rules than the real world does."

Rational minds operate on impressions they receive from limited perspectives, just as those imprisoned in Plato's cave do, to form structures which determine what they will and will not accept freely. This limited perspective is based upon human sensory information which perceives a three-dimensional coordinate system within tiny time and place references in the Universe.

Einstein discovered that such observational laws are valid only in certain regions of space where we reside. As our experience expands, we encounter more difficulty superimposing that reality upon other expanses where the rules no longer work.

General Relativity extends Einstein's Special Relativity theory, which applied only to coordinate systems moving uniformly relatively to each other, in order to examine phenomenon seen from two different frames of reference where one is moving uniformly and the other is moving non-uniformly. In other words, it describes events which occur in a coordinate system that is moving non-uniformly in terms which are meaningful to an observer in a coordinate system which is moving uniformly.

For example, imagine that the cable supporting an elevator in a very tall building snaps so that the compartment plummets downward. The people inside are unaware that this has happened because they can't look outside. According to Newton's Law of Gravity, the cabin accelerates as it falls...therefore its motion is not uniform.

Something strange happens within the cabin. Under normal conditions, an object dropped by someone, a pen or handkerchief for example, would fall towards the floor. Gravity would pull it there, just as it ordinarily holds the passengers to the floor. In this falling elevator case, however, they observe that the dropped objects won't fall, but simply hang in the air. If one were given a push, it would go straight forward until it hit an inside wall of the cabin.

To the observers inside the falling cabin who are experiencing an inertial coordinate system, there are no forces acting upon any objects around them. Nevertheless, rules of Newton's physics still apply...an object at rest remains at rest, an object in motion remains in motion, and for every reaction, there is an equal and opposite reaction. If one of the passengers pushed a handkerchief, the handkerchief would seem to push back...but with a force too small to be readily detected.

Sadly, the force they will experience when the elevator cabin reaches the bottom of its shaft is quite a different matter altogether. It is probably better that the clueless passengers are not aware of this abrupt and final end point.

Einstein concluded from such a thought experiment that gravity is equivalent to acceleration...a realization which led to his General Theory of Relativity. This entailed two ways of envisioning mass. One is gravitational mass, or the weight of an object that might be measured on a balance scale...the amount of force the Earth appears to tug on the object. The second type of mass is inertial mass...a measure of the resistance of an object to acceleration or deceleration (which is simply negative acceleration). This is why a pen and feather dropped in a vacuum with no air resistance would fall at exactly the same velocity.

Larry Bell

Einstein was not the first to discover that inertial mass and gravitational mass are equivalent; this was recognized three hundred years earlier, but it was disregarded as merely a coincidence. Einstein decided that it was just too much of a coincidence to be ignored.

He reasoned that his Special Theory of Relativity, which deals with un-accelerated (uniform) motion, was inadequate to describe much of what is going on in most of our Universe. Included are remote regions of space which are far from centers of gravity, as well as within very small subatomic regions of space...the domain of quantum mechanics.

Einstein visualized the Universe as a relatively flat plane which is curved by large masses such as celestial bodies which cause bumps. The larger the masses, the larger the flat space-time distortions. A mass the size of a star causes a relatively big bump, compared with orbiting planets which take the easiest paths around them.

Planets and other objects in our Solar System move as they do not because of gravity exerted upon them as a distance from the Sun, but rather because of the terrain of the neighborhood they are traveling through. English astronomer, physicist and mathematician Sir Arthur Stanley Eddington drew an analogy involving a big sunfish he whimsically, and perhaps appropriately, named Albert.

As we peer from our boat into the clear water, we observe that all smaller fish appear to be repelled from a point near the sandy lake bottom. They swim either to the left or right of it, but never over it. When we look closer, we notice that Albert has buried himself in the sand and has created a sizeable mound exactly at that spot. As the fish approach, they are simply following the easiest available path around Albert's mound.

Eddington offered this analogy to illustrate that if we could visualize Einstein's space-time continuum model, we would understand that planets do not move around the Sun because of "forces between them," but rather because of the geography (geometry) along their pathways. We can't directly see that geometry because it is four-dimensional, while our sensory experience is limited to only three dimensions.

Imagine life in a flat world of two-dimensional people who have only height and width and no concept of a third dimension. A straight line between them would appear as a wall which they can walk around but never step over in a mysterious (to them) third dimension. Spheres would not exist in their experiences...only circles. In a flat world, which they can only exist within—not on the surface of—two individuals setting off in opposite directions would never meet again.

Another way to help visualize influences of gravity (acceleration) upon space-time is to imagine a coordinate grid marked upon a stretched membrane of rubber with all lines straight, and all clocks synchronized. Wherever there is a large piece of matter such as the Sun, it distorts (warps) the space-time continuum membrane so that planets are deflected from straight line paths. Like the fish in Eddington's analogy, they take the easiest path around it.

Theoretical physicist Lisa Randall proposed that extra dimensions can be infinite in size yet unseen. This explains why parallel universes might be unseen with notably different properties than in our observed Universe. Imagining more than three spatial dimensions can be as dizzying as trying to imagine infinity.[16]

An analogy to unseen extra dimensions is illustrated by Teflon quasicrystals applied as a coating for non-stick frying pans. Whereas most crystals can be described by their 3-dimensional structure, quasicrystals do not conform to any of these usual patterns. Their strange arrangement of atoms precluding them binding to each other might similarly explain invisible connections in particle physics and cosmology.

Superstrings are a special class of hypotheses that consist of a ten-dimensional standard model: one dimension of time and nine dimensions of space. Rather than points, one-dimensional strings vibrating in 9-dimensional space are fundamental.[17]

Among unified concepts of elementary particle interaction, superstring includes gravity along with the other three fundamental forces. Broad inclusion led to the expression "theory of everything."

Like point matter and point anti-matter, strings and anti-strings annihilate each other. Superstring hypotheses contend that before the Big Bang, spatial dimensions existed at Planck length (about 10^{-33} centimeters). At this infinitesimal dimension, quantum fluctuations are extraordinary.

According to superstring hypotheses, the Universe existed 10-dimensionally in its pre-Big Bang form. Tied in intricate knots, the ten dimensions were ruled by quantum mechanics, fluctuating from tied to untied and back to different convoluted forms.

The Big Bang then created our three-spatial and unitary time dimension Universe. Remaining spatial dimensions became another universe, tightly coiled and infinitesimal. In the superstring concept, many other universes exist with more or less spatial dimensions than ours. String theory suggests our universe is one of 10^{100} possibilities most inhospitable.[18]

In 1983 Russian-American physicist Andrei Linde proposed the chaotic inflationary multiverse concept whereby random, chaotic quantum fluctuations cause new universes to emerge from previously existing ones rather than from a primeval Big Bang. Laws of physics and space-time dimensions might be different in each universe, with some similar to ours, and others bizarrely different. Following this logic, our Universe, which originated through a quantum fluctuation of another universe, might eventually fluctuate into a new and very different one.[19]

Edwin Hubble's discovery of time expansion in 1929 was extrapolated to a singularity, at which time the Universe was believed infinite in density and curvature. Theoretical physicists Stephen Hawking and Roger Penrose developed mathematical techniques that show if general relativity is correct, the Universe started as a singularity.

Others believe that quantum physics precludes the Universe from being a dimensionless point—one which is necessarily an infinitesimal with minute dimensions. Hawking subsequently observed that a theory combining general relativity and quantum mechanics is needed to accurately consider Universe origin— called 'quantum gravity.'

A five-dimensional Universe concept originally called a "big bounce" visualizes a contracting universe bouncing back just before collapsing in a big crunch into a tiny singularity.

Conciliation of general relativity and quantum physics has been offered recently with a Loop Quantum Gravity (LQG) theory which proposes that space-time consists of interconnected loops, each 10^{-35} meters in diameter. The loops appear smooth, similar to a cotton shirt made up of individual threads. Curvature of LQG creates gravity as described by general relativity.

Pennsylvania State University physicists Abhay Ashtekar, Tomasz Pawlowski and Parampreet Singh ran equations of LQG for the Universe backwards in time and arrived at a derivative concept that space shrinkage causes matter and energy compression.

In classical cosmology, Big Bang equations can only be reversed to a point beyond which classical laws break down. But at this point, LQG indicates extreme density tears down space-time fabric, the loops resisting further shrinking. Here, quantum gravity repulses, preventing further collapse and then the loops rearrange back to smooth fabric, allowing a universe to bounce back.

Einstein's General Theory of Relativity seems both beautifully elegant and enormously challenging to comprehend. On one hand, it informs us that the presence of mass/energy determines the geometry of space, and that simultaneously, the geometry of space determines the motion of mass/energy. Planets (chunks of matter) which had previously been regarded to be held in orbits around the Sun by gravitational attraction (force fields), are more realistically described as taking easiest paths through the curved space-time topography.

Yet just as that gravitational attraction exists only as the equivalent of motion, there is also no real thing as matter, which exists only as a curvature in the space-time continuum. There is really not even such a thing as energy, which only exists as mass operating in, and simultaneously influences space-time curvature. And to top it off, this can only really be described using very complex mathematics, where a fourth dimension—time—exists only in theoretical equations. Nonetheless, so far, it all works.

According to physicist and writer Richard Morris:

> *Quantum mechanics is an amazingly accurate theory, probably the most successful theory scientists ever created. It is the foundation of virtually all modern physics, and there is no known experimental evidence that contradicts it.*[20]

Minds over Matter

So, what if matter is only illusory after all? Does it really matter?

And if time is merely a non-linear contrivance—no previous *past* leading to a present *now* followed by a future and unknowable *when*—will this change normal routines such as having to get bills paid *on time*? Don't try that particularly dangerous experiment at home.

Interweaving time and space into space-time in a mathematical sense in accord with Einstein's general relativity, each component expands or contracts in mutual accord. Delete time, and no dimension exists in which an event can occur, and past events disappear. Delete space and time no longer exists, nor anything else.

And should we perhaps just reconcile ourselves to be satisfied to live in a three-dimensional world, or at least one that gives every outward appearance to our five senses of being one? After all, since that fourth dimension primarily exists in mathematical theorems and multi-world hypotheses, it will not change pragmatic realities and responsibilities which demand our single-world attention.

Sure, but is it not also interesting to contemplate how some of those subatomic mysteries connect with a really big Universe of metaphysical potentials which simultaneously exist in both the in here within each of us and the out there which we are all part of?

Consider, for example, that distinctions between organic (living) and inorganic (inanimate) stuff are meaningless at the subatomic level. As Gary Zukav points out:

> *Some biologists believe that a single plant cell carries within it the capability to reproduce the entire plant. Similarly, the philosophical implication of quantum mechanics is that all of the things in our Universe (including us) that appear to exist independently are actually parts of one all-encompassing organic pattern, and that no parts of that pattern are ever really separate from it or each other.*[21]

And how is it possible for those incredibly teensy particles and/or waves to instantaneously "know" what decisions are being made elsewhere and anywhere—then appear to "exist" here and there at the same space-time? Might this imply that being both here and there at the same time indicates that all particles—the ones that make up you and me very much included—are integrally connected within an in-here/out-there Universe of Everywhere?

Is there such a thing as a conscious Universe which cannot be consciously accessed through our five senses alone? Physicist E.H. Walker has speculated that even photons may be conscious:

> *Consciousness may be associated with all quantum processes...since everything that occurs is ultimately the result of one or more quantum mechanical events, the Universe is 'inhabited' by an almost unlimited number of rather discrete conscious, usually unthinking entities that are responsible for the detailed working of the Universe.*[22]

Although Stephen Hawking and Albert Einstein expressed skeptical views regarding God and religion, both delved deeply into mysterious workings of nature at a subatomic level which, to our conventional senses, take on extrasensory, supernatural manifestations.

Quantum mechanics theories go even so far as to suggest a preposterous possibility that everything in the physical Universe exists only as illusory inventions of our individual minds. Whereas this concept presents a radical departure from traditional Western thought, it does not seem nearly so alien to much older Eastern philosophies.

Generally speaking, whereas Western philosophies tend to emphasize learning new things about what reality is, ancient Hindu and Buddhist literature speaks of removing veils of ignorance that stand between us and what we really are. And where Western religions tend to envision a Universe divided into separate material and spiritual aspects, Eastern teachings make no dichotomous distinctions between material and spiritual manifestations.

Quantum mechanics challenges any notion of material reality altogether, making no distinction between mass (quanta) and their energetic and mysteriously unpredictable relationships with individual observers. In doing so, it has yielded replicable evidence that powers of mind over matter, and realities much stranger than presumed fictions, can no longer be casually dismissed merely as quack clichés.

As Gary Zukav observes in his illuminating book *The Dancing Wu Li Masters: An Overview of the New Physics,* the new physics shakes foundations of the old physics. The old physics has assumed that there is an external world out there which exists apart from the I which is in here. This I within each of us can observe and measure the world without changing it.[23]

Zukav urges us to reconsider the old, traditional way of thinking which emphasizes that an absolute truth exists out there which we become closer to understanding each time our scientific approximations become closer to explaining it. Such explanations do not necessarily reflect the way things (or phenomenon) really are…or as Einstein put it, "We still will not be able to open the watch."

Quantum theory delves bravely into such ideas, while simultaneously suggesting that all ideas regarding absolute truths are illusory. As Gary Zukav explains, "All attempts to describe 'reality' are relegated to the realm of metaphysical speculation."

Zukav describes this scientific and mental paradox:

> *The mind is such that it deals only with ideas. It is not possible for the mind to relate to anything other than ideas. Therefore, it is not correct to think that the mind actually can ponder reality. All the mind can ponder is its ideas about reality. (Whether or not that is the way reality actually is, is a metaphysical issue.) Therefore, whether*

Larry Bell

*or not something is true is not a matter of how
closely it corresponds to the absolute truth, but
of how consistent it is with our experience.*[24]

Henry Pierce Stapp summed up the foundational philosophy,
saying:

> *[Quantum mechanics] was essentially a rejection
> of the presumption that nature could be
> understood in terms of elementary space-time
> realities. According to the new view, the
> complete description of nature at the atomic
> level was given by probability functions that
> referred, not to understanding microscopic
> space-time realities, but rather to the
> macroscopic objects of sense experience. The
> theoretical structure did not extend down and
> anchor itself on fundamental space-time
> realities. Instead, it turned back and anchored
> itself in the concrete sense realities that form
> the basis of social life...This pragmatic
> description is to be contrasted with descriptions
> that attempt to peer 'behind the scenes' and tell
> us what is 'really happening.'*[25]

As science peers behind those scenes, a growing body of evidence
indicates that the distinction between the in here and the out
there is an illusion.

As Gary Zukav observes:

> *Now, after three centuries, the Scientists have
> returned with their discoveries. They are as
> perplexed as we are (those who have given
> thought to what is happening). 'We are not*

32

sure,' they tell us, 'but we have accumulated evidence which indicates that the key to understanding the Universe is you.'

Zukav concludes: "If the new physics has led us anywhere, it is back to ourselves."[26]

Order Emerging from Chaos

Quantum mechanics has taught us that the vacuum of space-time is not as empty as previously thought, but rather, is alive with subatomic particles and antiparticles in continuous creation and annihilation. Here, an electron is an example of matter, a positron its antimatter. It is theorized that quantum fluctuation in subatomic space-time foam may have served as the primordial Universe's emergent mechanism.[27]

According to Big Bang theory, a tiny fraction of a second after a condition of infinite temperature the Universe temperature dropped to about 10^{15} degrees where physical theories first become meaningful.

Although originally symmetrical like a crystal, cooling caused this initial super force to split into four forces: the weak force associated with radioactive decay, the strong force that holds the nucleus of atoms together, the electromagnetic force that is the attraction or repulsion between electrically charged particles and the gravitational force. Asymmetry became the basis of cosmic structures and life.

Viewing these from a functional perspective, the strong force that binds quarks into neutrons and protons produces energy from nuclear explosions; the electromagnetic force dominates biological processes; the weak force permits neutrons to decay, for example from Uranium, into less weighty elements and gravity which pervades around all matter and increases with increasing mass.

Larry Bell

At short distances, gravity is the weakest force, but at greater distances gravity dominates.

After the Big Bang, a primordial plasma of subatomic particles appeared: quarks, electrons, neutrinos and protons, which combined to form atomic nuclei. With cooling, electrons combined with atomic nuclei to form hydrogen and helium atoms along with traces of lithium and deuterium. Hydrogen is the most abundant Universe element.[28]

Expansion and further cooling permitted slight excess of matter over antimatter, some subatomic matter changing into light atomic nuclei three minutes later.

Physicist Alan Guth proposed an inflation theory which concluded that the within 10^{-35} seconds of age the primordial Universe had expanded faster than the speed of light by an astounding factor of nearly 10^{50}! Surprisingly while Einstein's relativity shows the speed of light (photons) is the speed limit of matter, spatial expansion is not restricted by this.

Since confirmed by the NASA's Cosmic Background Explorer (COBE) satellite, Guth believes that the Universe continues to expand at a more moderate degree today.[29, 30]

With a bang, universes expand in accord with their density. If density exists at a critical level, a universe will expand forever. If density exists less than critical, hydrogen and helium would disperse too thinly to form cosmic bodies. If density exists greater than critical level, universes might contract in a big crunch.[31]

Dark vacuum energy is a strange product of relativistic gravity, out-competing gravity to affect accelerated Universe expansion—the same as Einstein's cosmological constant. Because it is extremely small-scale, dark vacuum energy exists as a quantum system, wavy and uncertain.

Four hundred thousand years after the bang, the primordial plasma thinned and became transparent. Set free, quantum particles of light called photons made the Universe visible.

Cosmic Microwave Background (CMB) consists of photons that permeated space since the short time after the Big Bang.

CMB temperatures have been calculated at three degrees above absolute zero. Cosmologists detect microwave radiation by radio telescopes; others may observe CMB by snow on TV screens.

Light waves or subatomic waves generally might be the collective behavior of dense swarms of quantum particles. If so, this phenomenon illustrates how collective behavior in both quantum and Newtonian worlds can be greater than the sum of the parts. Never quiet, quantum systems appear to exist as perpetual motion machines.

Strangely, at these minute scales quantum photon particles behave both as waves and particles. Molecular geneticist Johnjoe McFadden analogized light waves with water waves, which are the collective dynamics of a myriad of water molecules.

Using the Relativistic Heavy Ion Collider at the Brookhaven National Laboratory, physicists Michael Riordan and William Zajc affected a mini big bang. Gold nuclei traveling at near light speed were collided to effect quark gluon plasma. Surprisingly, the plasma did not behave like a gas, but instead like a nearly perfect liquid.

Quantum waves have a probability of existing in multiple places simultaneously, perhaps existing in our Universe and other Universes at once. Known as tunneling, the waves penetrate seemingly impenetrable barriers.

Most Universe material does not emit radiation and is called dark matter. Only about four percent consists of ordinary atoms called baryonic, the remainder exotic matter and energy.[32]

Einstein's general relativity implies our Universe may have been created by a massive black hole mechanism. It is theorized that when stars eventually become supernovas or gamma-ray bursts, some of their material collapses into black holes. A portion of that matter that falls through a black hole then travels through a wormhole into a newly created world of space-time. If this baby universe experiences inflation, a new large universe emerges.[33]

Molecular clouds are believed to have first collapsed to form black holes; only then did galaxy stars form. Billions of years later, greater heat and pressure in star interiors forged molecular gases.

Matter in the Universe became uniformly distributed on a large scale, but clumpy on smaller scales. Beginning early, matter clumped into dense regions, possibly originating as quantum fluctuations, galaxies then forming by gravitational action. Physicist Fred Adams offered the example of a flickering candle to roughly visualize quantum flux. The flame continuously occupies different space, more in some places, less in others.[34]

Although black holes are scattered throughout the Universe, huge black holes seemingly exist at the core of all galaxies. They are so dense and the gravitational attraction is so great that even light cannot escape their grasp.

The primordial Universe is believed to have been similar but of far greater density than black holes today. Still outside the singularity, time freezes at a black hole's "event horizon," within which there is no escape. With loss of its time associate at the event horizon, space atomizes and creates foam from which new universes might continuously emerge. With the emergence of a new universe, space and time re-unite.

Life and light on Earth would be nil without star energy, which slows total Universe disorder—the ultimate fate of closed systems without influx of energy. Star energy develops from four hydrogen atoms fusing to form helium. Four hydrogen atoms have greater mass than helium, the difference representing energy conservation.

Exploding red giant stars, supernovas and gamma-ray bursts create inordinately high temperatures and pressures which broadcast space with clouds of debris enriched with heavy elements. Our Sun will become a red giant in some 5-7 billion years, but it is not large enough to become a supernova.

Exhaustion of a star's nuclear fuel causes the residual iron core to collapse on itself. When stars having between 10 and 20 times the mass of the Sun implode as supernovas, part of their mass collapses to form extremely dense neutron stars that scatter vast extents of space with interstellar dust. If larger than 20 times the mass of the Sun, the star implodes as a hypernova to form even denser bodies or black holes.

Billions of years later, other stars and planets are formed from interstellar dust derived from imploding stars. The heavy material elements were essential to rocky planet formation, including Earth. Carbon and other life-essential elements are also produced by star-interior and implosion extremes of temperatures and pressures.

Looking back with telescopes to observe how the early Universe appeared shortly after the Big Bang, our perspective is limited to only a minute portion of our bubble or perhaps multiverse. Nevertheless, what we view reveals a fascinating scene of constant change dramatically punctuated by supernova explosions and gamma-ray fireworks.

Altogether, we can be very grateful that expansion and cooling of the primordial Universe led to a fortuitous sequence which created subatomic particles, atomic nuclei, atoms, galaxies, stars, molecules, some heavy elements, imploding stars, more heavy elements, terrestrial planets and ultimately—some 10 billion years later after Earth's viable environment emerged—life.

As noted by Wayne Bundy, the fundamental changes in the Universe can be compared with the essential quality of plasticity in our human brains.

He writes:

Both the Universe and our brains gain

knowledge by structural devolution and evolution. If not for continuous change, knowledge would not evolve in the Universe or the human mind. Time would freeze! Information's evolutionary end would spell the end of biological evolution and life.[35]

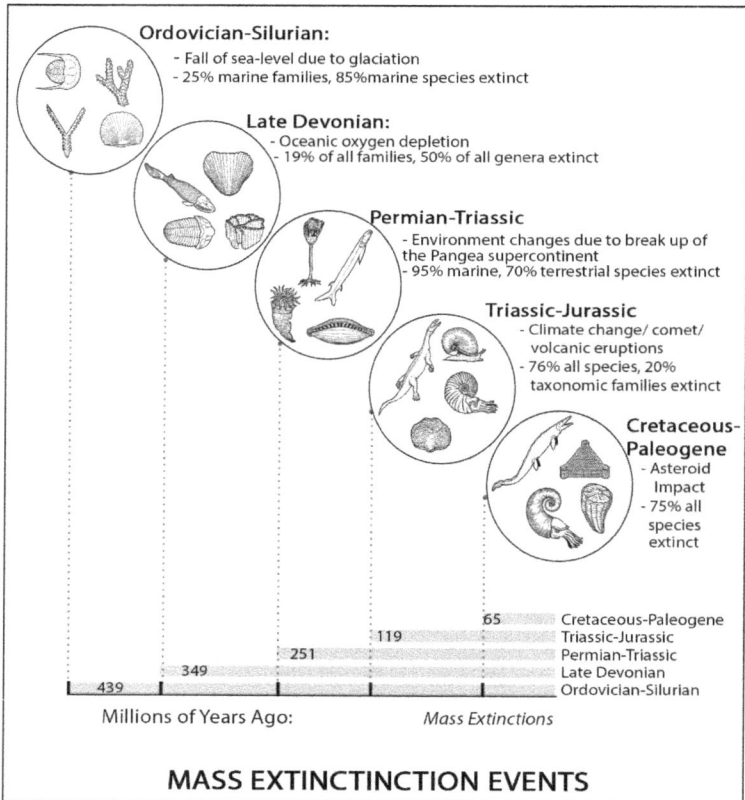

Ordovician-Silurian:
- Fall of sea-level due to glaciation
- 25% marine families, 85%marine species extinct

Late Devonian:
- Oceanic oxygen depletion
- 19% of all families, 50% of all genera extinct

Permian-Triassic
- Environment changes due to break up of the Pangea supercontinent
- 95% marine, 70% terrestrial species extinct

Triassic-Jurassic
- Climate change/ comet/ volcanic eruptions
- 76% all species, 20% taxonomic families extinct

Cretaceous-Paleogene
- Asteroid Impact
- 75% all species extinct

65	Cretaceous-Paleogene
119	Triassic-Jurassic
251	Permian-Triassic
349	Late Devonian
439	Ordovician-Silurian

Millions of Years Ago: *Mass Extinctions*

MASS EXTINCTINCTION EVENTS

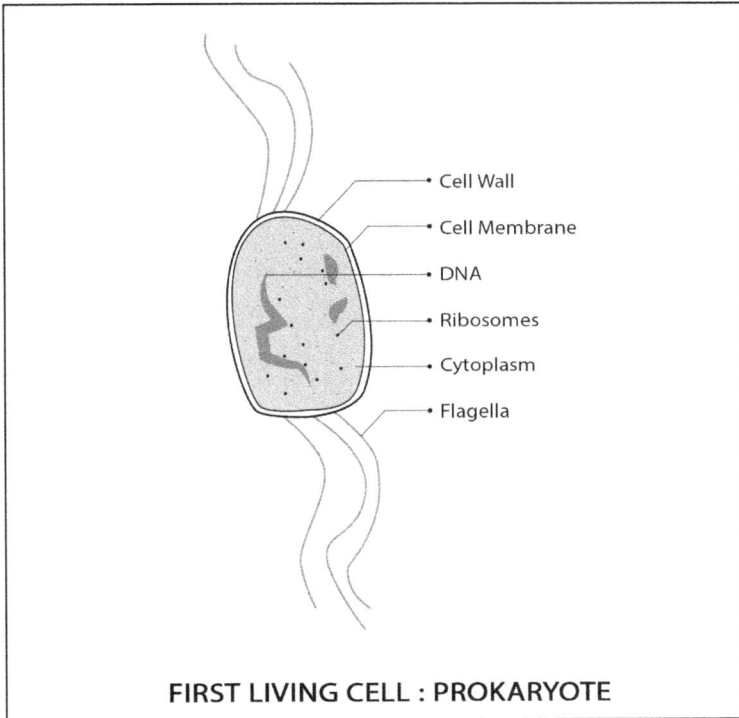

FIRST LIVING CELL : PROKARYOTE

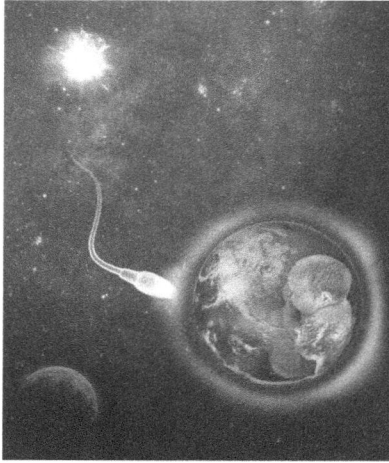

Part Two: The Emergence of Life

IT IS IMPOSSIBLE to even begin to comprehend the staggeringly statistical odds against life existing on the particular planet we have the marvelous good fortune to inhabit, our own physically and intellectually complex lives very much in particular. But then again, if not for that myriad of incomprehensibly unlikely events which happened to occur at exactly the right places and times, we would not have any reason or capacity to wonder at all.

Borrowing a much quoted *cogito ergo sum* phrase, "I think, therefore I am," Rene Descartes noted in his *Principles of Philosophy:*

> *While we thus reject all of which we can entertain the smallest doubt, and even imagine that it is false, we easily indeed suppose that there is neither God, nor sky, nor bodies, and that we ourselves even have neither hands nor feet, nor, finally, a body; but we cannot in the*

same way suppose that we are not while we doubt of the truth of these things; for there is a repugnance in conceiving that what thinks does not exist at the very time when it thinks.

So now that we are graced with an ability to think about it, was Earth's habitable development which led to our human existence an unbelievably lucky draw in a cosmic lottery? One of my previous books, *Cosmic Musings: Contemplating Life Beyond Self,* ponders this fundamental question from the viewpoints of diverse religious, philosophical and scientific scholars throughout social history. Each person who is compelled to wonder about such matters must seek their own personal answers.

This book makes no pretext of addressing the ultimate question of the *whys* behind these marvels, but only to summarize what current scientific thought tends to inform us about some of the *hows*...the mechanisms of inevitability, chance or both that brought them about. Such integrated fields of confluence include cosmology, meteorology, physics, chemistry, geology, paleontology and paleoanthropology to mention but a few.

To begin, whether by realm of preordained destiny or unpredictable quantum flux, Earth-type planets capable of supporting complex life are statistically rare beyond comprehension. For whatever reasons, our celestial home came about in exactly the perfect neighborhood, with remarkable timing and exactly the right resources and environmental conditions.

The Makings of a Habitable Planet

Assuming a Big Bang of a beginning, the primordial Milky Way galaxy arose from a gas cloud which originated from the cooling of dark matter many billions of years after time's beginning.

Then, about 4.7 billion years ago, a mass of gas and dust detached from a spiral arm of the galaxy and was formed into a disk through angular momentum. Our Sun condensed from that disk.[36]

Another theory holds that our spiral, disc-shaped Milky Way galaxy was formed through a merger of two elliptical galaxies. Such reasoning suggests that matter is now too thinly dispersed for new galaxies to form, causing ongoing galaxy growth through assimilations with satellite galaxies and inter-galactic gas clouds.[37]

Planets began as microscopic grains of dust and grew by actions of electrostatic forces. Impacts and abrasions between these particles randomly dislodged electrons, causing some to be more electropositive and some more electronegative. Continuous aggregation occurred through oppositely charged particles. Orbital speed differences led to the build-up of even larger sizes, and gravitational attraction determined final planet mass.

The primitive, cold nebula constituent materials consisted of about 98 percent hydrogen and helium, and only about 2 percent heavier elements. For this reason, rocky or terrestrial planets such as Earth emerged relatively smaller in size.[38]

Because storming plasma jets from stars erode gas from planets in their accumulation, rapid formation and increasing distance from the Sun favors gaseous planets. The huge size of Jupiter, for example, may largely account for the small sizes of Earth and Mars.

Complex life would not have evolved without both continuing change as well as stability. Had our Homo sapiens ancestors arrived too early, suitable conditions for emergence and survival would not have been ready for them. If those essential conditions had not been sufficiently stable over the past 150,000 years, there would not have been enough time to reach that stage of being "us."

The Importance of Location, Location, Location

Although an enormous black hole at the Milky Way center helps to stabilize the galaxy, it exerts inadequate gravitational attraction to hold everything together. Dark matter surrounding the galaxy does the lion's share of stabilization by its enormous gravitational strength. Nevertheless, our solar system developed at a fortuitously uncrowded orbital location which minimized disruptive and chaotic gravitational orbit-altering influences of other neighboring bodies upon its continuity.[39]

The potentially habitable real estate zone necessary to allow liquid water and metals for stable terrestrial planet formation within the spiral Milky Way galaxy is relatively narrow, extending from about two-thirds the distance from its center to periphery. The cosmic science jargon for metals includes all elements heavier than helium. Toward the galactic core, density and energetics become excessive for stable terrestrial planets. Moving outward from the habitable zone, too few metals exist for terrestrial planet formation.[40]

Earth's habitable zone from the Sun is critical also to enable liquid water along with other vital conditions such as an oxygen atmosphere essential for complex life forms to exist. This region extends within a distance of 0.95 to 1.15 astronomical units (93 million miles per unit) around the Sun.

Although other atmospheres could support a lesser array of life forms, such as anaerobic bacteria, viable temperature range for Earth's aerobic life requires optimum levels of greenhouse gases containing carbon.[41]

Plasma jets emitted from the Sun stormed the early solar system with gale-force solar winds which swept away much of the light gases from inner planets. The outer planets were affected far less, which accounts for their high ice-gas nature.

Beyond critical size, planets hold an atmosphere. If too big, excessive pressure and temperature break inter-atomic and molecular bonds of matter. Structures in the Universe stabilize by a balance between opposing forces of nature. For example, a balance between gravitational and atomic forces exists when the matter's density is near that of the interior of a single atom.[42]

A Most Favorable Sun

Among many fortunes allowing life on Earth, by astronomically good luck, both our Sun and planetary abode developed nearly optimum masses. The Sun is also a proper size to extend very long survival.

The Sun curves space in accord with its large size and mass. Planets follow this curvature of space, the geodesic, which can be compared to walking a contour around a depressed area. Gravitational space distortion by neighboring planets precludes perfect orbital symmetry.[43]

Constant burning of Universe energy is the single most important ingredient that enables our existence. Earth continuously fuels environmental changes by utilizing energy from the Sun and by burning radioactive fuel in its interior essential for continuous geological and biological evolution.[44]

Whereas much smaller and dimmer stars occur far more commonly than our larger Sun, their meager outputs of energy would require life-supporting planets to orbit much closer than even Mercury does to our Sun. Whether such close life-sustaining orbits are feasible remains unknown. Unlike very high-mass stars which may survive as little as 300 years—not long enough for the likelihood of life to emerge—thrifty burning of hydrogen by small stars does offer the potential advantage of existing over trillions of years. Massive blue giants survive as little as 3 million years.[45]

When small stars exhaust their hydrogen, they become red dwarfs rather than red giants. As they decrease in size, they become hotter, brighter and bluer. Earth's yellow dwarf Sun, on the other hand, will probably become a red giant which grows to engulf our planet within another 5 billion to 7 billion years.

Spectacularly Fortunate Collisions

Colossal early collisions with other space bodies brought about enormously beneficial consequences that gave rise to our living planet. One of these, a real doozy involving a collision with a former planet posthumously named Theia which occurred during the final stages of Earth's accretion, led to the Moon's creation.

Theia is estimated to have been perhaps half the size of Mars. Its impact is credited with accounting for Earth's 24-hour rotation and tilted 23-degree axial ecliptic obliquity which influences seasonal climate changes.

Large cosmic impacts may also be responsible for causing the clockwise spin of Venus and Pluto along with Uranus's axial tilt of over 90 degrees. Viewed from the north, the rest of the planets spin counterclockwise.[46]

Earth's precession of the equinoxes in its orbit around the Sun is influenced with respect to the Moon's distance. Exerting a stronger pull on the closest side of Earth, the Moon also causes monthly tidal variation. This cycle affects many aspects of life, exemplified by the calendar and menstrual cycles of many mammals.

Beginning about 3.9 billion years ago when oceans first formed, the closeness of Moon to Earth caused tides nearly 10,000 feet high. During the Moon's perigee (orbit closest to Earth), tides continue to be at their highest today.

A total eclipse of the Moon aroused great fear and brought end to war between the Lydians and the Medes in 585 BC—and also between Athens and Syracuse in the Peloponnesian war in 413 BC. Yet within 500 million years the Moon's distance from Earth will be too great for a total Sun eclipse.[47]

Collisions with large objects have also influenced planetary orbits and ecliptic tilts of other planets with regard to their axial planes while orbiting the Sun. Earth's orbit is just slightly elliptical. Drastic climate changes would preclude complex life if Earth possessed a pronounced elliptical orbit.

Terrestrial planets have developed approximate elliptical orbits, while much larger Jovian planets, including Jupiter, Saturn, Uranus and Neptune, have developed approximate circular orbits. These outer planets furthest from the Sun are comprised of rocky cores enshrouded in ice and cold gas.[48]

The frequency of cosmic impacts decreases in proportion to debris size increases according to a well-established statistical power law. This ten-time increase factor in size corresponds to a rate of impact one hundred times less frequent.

As described by physicist Fred Adams, these mathematical relationships suggest an underlying principle of natural order. Whereas a one-meter objects hit the Earth yearly, one-kilometer objects can be expected to impact the Earth on an average of approximately only once every million years.[49]

Subject to violent cosmological events, life on Earth is not safe from a catastrophic meteorite strike of the scale that exterminated dinosaurs. But by the same token, we can be grateful that life has also become much safer from an asteroid belt between Mars and Jupiter by a big reduction in cosmic debris vacuumed by Jupiter.[50]

Between Neptune and Pluto are icy bodies known as trans-Neptune bodies or Kuiper-belt objects. Beyond Pluto (which was recently booted from planet classification due to small size) comets made of ice and rocky debris comprise an Oort cloud.

Many left-over bits of rock and ice have continued to bombard the planets for billions of years after the Big Bang. Erosion destroyed most of the impact craters formed on Earth, but Moon's non-atmosphere allowed freedom from weathering and erosion, enabling preservation of impact craters.

Opportune Life Timing

An extreme spin-rate made hurricane force winds commonplace when Earth's first formed. Those enormously strong winds along with high temperatures and atmospheric oxygen's absence would have precluded most life until it first emerged nearly 3 billion years after the Big Bang. The spin also energized a strong geodynamo magnetic field around Earth which continues to deflect deadly cosmic radiation.

The geodynamo was made possible as the result of innumerable impacts during primordial Earth's formation which produced a molten condition that caused heavier surface elements, (iron and nickel in particular) to slowly sink to its core. The Earth's spin and convection currents produced by slow circulating motions in the molten iron-nickel core created a huge electromagnetic generator.[51]

Without molten cores, the Moon and Mars do not have a magnetic field. If the Earth were to rotate faster, as it did in early history, the magnetic field would be much stronger. Over thousands or millions of years the Earth's magnetic field reverses. Why this happens is one of the great mysteries of Earth's behavior.

The Earth's magnetic field also prevents solar winds from casting our atmosphere required for life into space by a process called "sputtering." Absent oxygen required for the formation of an ozone layer, lethal ultraviolet radiation in addition to cosmic radiation would have destroyed most early emerging land life. A stronger geomagnetic field offered by early Earth's faster rotation conceivably would have provided even more radiation protection.

Close to the geographic pole, the magnetic North Pole is unstable and migrates slightly over time. Such migration may result, for example, from a super-volcanic eruption requiring adjustment of Earth's weight distribution.

Larry Bell

The magnetic pole decreases in intensity over time as well, with a loss of about five percent in the last one hundred years. This decrease in magnitude relates to Earth's cooling interior as a result of slowly depleting radioactivity.

Also powered by energetics of Earth's molten core, our planet's dynamic plate tectonics architecture acts to moderate climate and to form and reformulate continents and coastal areas where land and sea life flourishes. Without plate tectonics, there would be no plant growth carbon cycle, and mountains would eventually erode away except for an occasional volcano.

Sometimes called continental drift, plate tectonics involves the continuous displacement of the Earth's thin crust over the mantel surrounding the molten core which is kept hot by the gradual decay of radioactive materials. Convection currents which develop in the molten core and mantle cause hot materials to rise toward the surface as cooler materials descend.

Sluggish but continuous movement of the mantle drags against the crust with enormous resistance, fractures it into plates and slowly displaces them horizontally by roughly one inch per year.

Although this may seem small, the movements of larger plates over geologic time can produce a continent or major part of an ocean basin. As plates collide with others, one more may descend back into the mantle (called subduction). In other boundary instances the plates pull apart.

Plate contacts are areas of weakness which provide escape paths for volcanic eruptions which emit carbon dioxide and acidic debris. The Ring of Fire at the Pacific Ocean's perimeter is a major example. Associated with mid-ocean ridges, volcanic vents develop between both retreating and abutting plates.

Carbon dioxide and other greenhouse gases (gases containing more than two carbon atoms per molecule) must exist within a reasonable range to maintain viable climates. The Sun's radiation (relatively short wavelengths) is absorbed by Earth.

This absorbed radiation is transformed and reflected from Earth's surface as infrared wavelength, which is heat.[52]

CO_2 is also vital to plant growth. Green leaves use energy from sunlight through photosynthesis to chemically combine carbon dioxide from the air and water with nutrients trapped from the ground to produce sugars, which are the main source of food, fuel and fiber for life on Earth.

Even inactive volcanoes constantly leak carbon dioxide. The atmospheric CO_2 build-up is moderated over time by a reaction with silicate rocks which led to the first stages of a limestone formation on the ocean floor.

Then, after life formed, additional mechanisms for ocean carbon storage were provided by plant and algal growth and the burial of organic matter.

Diatoms and radiolarian (algae) use silicic acid to make their silica shells. Coccolithophores (algae) convert calcium bicarbonate into calcium carbonate shells. With algal death, calcium carbonate ($CaCO$) shells deposit on the ocean floor, leading to more limestone formation. Similarly, silica shells deposit to form diatomaceous earth.

Together with early plant materials, nature has generously provided all the building blocks necessary to evolve complex organisms...us included.

Extraterrestrial sources of water and amino acids which occur in complex carbon molecules have arrived via a diverse spectra of high impact comet and meteorite deliveries.

These, in turn, originated in giant molecular clouds in the Milky Way which contain organic materials as well. Included are free atoms of hydrogen and helium and a wide variety of carbon-containing molecules such as carbon monoxide, formaldehyde and alcohol.

Under appropriate conditions these organic materials react to form amino acids—essential components for proteins of life.[53]

Life's Early Beginnings

Before we consider how "life" began, let's contemplate how that first living evidence might be recognized as the "real thing." Here, life's fundamental definition remains an enigma, wherein not every "expert" will agree.

Earth scientist Robert Hazen at the Carnegie Institution of Washington's Geophysical Laboratory has proposed that any definition of first life is doomed because the transition from non-life advances stepwise toward increasing complexity. He argues that rather than perceiving a rigid dichotomy that divides the natural world of living versus non-living, life arose in a gradual sequence of emergent steps "from geochemical simplicity to biological complexity." [54]

Biology and biotechnology Professor Lee M. Silver at the George Church at Harvard Medical School theorized that life can be created with the correct combination of off-the-shelf chemicals. The idea is to synthesize sentient, replicating organisms capable of undergoing at least one thermodynamic cycle that can eventually evolve to other life forms. [55]

Creating a long-lived, energy using, self-replicating, metabolizing organic system from scratch with simple chemicals has been a long-sought goal. As elaborated by Silver, this extraordinary biological act was last performed by nature on Earth almost 4 billion years ago.

First life may have been a self-replicating, catalytic RNA (ribonucleic acid) module in an ocean that contains codes for protein assembly. Single-cell bacteria and archaeans which emerged later combined the RNA with DNA (deoxyribonucleic acid), the informational basis for life.

RNA is a single-strand chromosome, whereas DNA has a double helix structure, two chromosomes containing genes that spiral around each other. Like DNA, RNA contains four bases, adenine, uracil, guanine and cytosine, but in DNA, the uracil of RNA is replaced by thymine. Bonds between the bases create a ladder structure.

Simple life may have emerged numerous times in Earth's early history, for example, following repeated annihilations of predecessors by asteroid impacts. If so, regardless of how many times it arose, it seems always to have appeared with a common and essential amino acid base found in Milky Way galaxy cloud chemicals. This commonality might suggest that other life that may exist on other planets might use amino acid building blocks also.

Nucleotides—the structural units of nucleic acids that make up RNA and DNA—are fundamental to all life on Earth. Viruses which are sometimes characterized as existing at the boundary of non-life and life are DNA or RNA packaged in a protein shell and exist outside living cells. With cell invasion, some viruses cause damage, while others can be entirely beneficial. Either way, they often leave some of their genes behind or take some of the cell's genes with them when they vacate.

Viruses aren't generally considered to be alive in the sense of self-replication because they reproduce only by parasitizing their cell hosts. Geologist-biologist Peter Ward suggests that some viruses are underrated in this exclusion from life status. He points out that a great variety of genome variations which exist in viruses likely include some non-parasitic viruses—transitional forms in between viruses and bacteria.[56, 57]

Ward observes that viruses which are mostly much smaller than bacteria emerged in myriad varieties, especially in oceans. While bacteria have been hailed the most abundant life form, it is now known that there are one or two orders of magnitude more viruses than bacteria. One milliliter of seawater may contain as many as 10 million of them.

Viruses exist virtually everywhere, at the sea bottom and below, high in the atmosphere and inside every life form ever studied. Some are ancient, degenerate RNA organisms that preceded cellular life. Some have formed essential links for all forms in the tree of life. Still others originated parasitically, emerging after cellular life provided vulnerable hosts.

Single-cell organisms, including bacteria, Archaea and Eukarya, are hosts to a myriad of viruses, constantly changing to keep pace with their intruders' evolution. Bacteria and Archaea containing both RNA and DNA are the most metabolically diverse and smallest life forms that reproduce by dividing. Some, as well as certain protists and fungi, reproduce asexually. Dividing as clones, they remain forever young.

Others transfer DNA genes horizontally between one living cell to another. Such genetic DNA swapping between single-celled organisms became instrumental in plant and animal evolution.

Significantly departing from bacteria, archaebacterial became designated Archaea. Together, bacteria and Archaea make up more than half of the biota, the only life forms for the first three billion years of life's history. We can be grateful that enormous bacteria populations in our bodies help us to digest food, generate certain vitamins and restrain the growth of harmful bacteria.

Eukarya, which may have recently emerged about 1.7 billion years ago in Earth's 4.6-billion-year history, ate neither animals nor plants. They include ciliates, amoebae, malarial parasites, slime molds, plankton and seaweeds. One unicellular microbial form known as Euglena gracilis, contains photosynthesizing green parts like plants, yet swims animal-like.

Single-celled organisms throughout history very gradually helped to transform Earth's atmosphere into a composition that enables our existence today. They accomplished this by sustaining life-essential processes: fermentation, photosynthesis, oxygen breathing and fixing of atmospheric nitrogen into proteins.

The development of complex life depended upon

continuous resupply of viable oxygenated atmospheric composition. And while oxygen is highly toxic, aerobic organisms have a protective shell of enzymes that destroy such harmful chemicals as free radicals and hydrogen peroxide produced by oxygen reacting with tissues. Fortunately, we adapted too—our intelligent brains require lots of oxygen-derived energy.[58]

The emergence of complex life also depended upon symbiogenesis among unicellular and multi-cellular forms. Eukaryotes established a basis of multi-cellular life through a symbiotic relationship with mitochondria (which are parasites in cells) that burn oxygen for energy. This could not occur before atmospheric oxygen became abundant.

Biologist Lynn Margulus proposed in the mid-1960s that eukaryotic cells and their nucleus emerged much earlier—about 3 billion years ago—a development which occurred through an endosymbiotic merger between two kinds of single-celled life. Each of the two cells provided a survival benefit to the other—a large bacterium engulfing a smaller one, the latter becoming the nucleus.[59]

Margulis and writer Dorion Sagan suggested that whip-lashing spirochetes may have forged a symbiotic merger with eukaryotes, leading to cellular motion: "a sort of ménage à trois among thermoplasma, mitochondria and spirochetes."

First life may have been constructed on mineral surface templates which served as catalysts for chain formation of nucleic acids and amino acids. Chemist Graham Cairns-Smith proposed in the 1960s that clay materials may have served as crude enzyme catalysts as well, leading to the precursors of today's genes.

Strong acid surface layers and edges of plate-shaped clay materials are conducive to catalytic alteration of organic molecules. These clay materials are most likely to develop from basaltic rocks that enclose black smoker ocean vents.[60]

Taken altogether then, these theories suggest that the precursor living cell did not appear fully formed. Rather it very likely arose through a series of events—organic synthesis, molecular selection and encapsulation into diverse molecular structures. Yet the biggest question of all—what life really is, remains an unresolved semantic dispute of enormous complexity.

Ancient Lessons from Flowers and Fossils

We all tend to readily distinguish between animals and plants. Plants, after all, are generally immobile, rooted in the ground, spread their green leaves to heavens and feed on sunlight and soil. Animals, on the other hand, are mobile, forage or hunt for food. Plants and animals also evolved along two profoundly different paths (fungi yet another).

Until rather recently it has been considered sufficient to divide complex life forms into a strict two-kingdom taxonomy classification system, plants and animals. At least this was the case before English naturalist Charles Darwin came along and suggested that plants and animals are more closely related than generally imagined. Insect-eating plants, for example, use electrical currents to move just as animals do.

Yes, there are dramatic speed differences. Whereas plant electricity is much slower, roughly an inch a second, animal electricity conducted by nerves, moves roughly a thousand times faster.[61]

All signaling between cells depends on electrochemical changes, the flow of electrically charged atoms in and out of cells via special, highly selective molecular pores or channels. These ions cause electrical currents, impulses—action potentials—that are transmitted (directly or indirectly) in both plants and animals.

Animal speed depends upon sodium and potassium ion and channels that can open and close in a matter of milliseconds, allowing hundreds of potentials to be generated in a second. This evolutionary neuromodulation response which takes place at synapses has made it possible for some advanced organisms to learn, profit by experience, judge, act and finally think.

Although *Origin of the Species* is Darwin's most famous book, his own personal favorite titled *Fertilization of Orchids* published in 1862 was about co-dependent evolutionary relationships between insects and plants. This is not to suggest that Darwin wasn't just as much—perhaps more—interested in the evolution of animals, ultimately leading to humans. It was more that he recognized that any suggestion that Homo sapiens evolved from apes would invite hostile recriminations from prevailing creationist religion doctrinaires.

Nevertheless, while Darwin was very careful to say little in the book about human evolution, the implications of his theory were perfectly clear. He wrote to his friend Asa Gray, broadly considered to be the most important American 19th century botanist who argued that religion and science were not necessarily mutually exclusive, and said:

> *[No] one else has perceived that my chief interest in my orchid book, has been that it was a 'flank movement' on the [opposition] enemy.*[62]

Whereas the idea that man could be regarded as a mere animal—an ape—descended from other animals had provoked outrage and ridicule, plant evolution for most people was seen as a different matter. Plant evolution was perceived as less relevant and threatening to most people than the notion of animal evolution. Plants neither moved nor felt. They inhabited a kingdom of their own, separated by a great ideological gulf from the animal kingdom.

Swedish scientist Carl Linnaeus had shown in the eighteenth century that flowers had sexual organs (pistils and stamens). It was universally believed at the time that they were therefore self-fertilized. If not, they wondered, why would each flower contain both male and female organs?

Darwin made a radical break with conventional thinking by observing that flowers cross fertilized with other flowers through "contrivances" that co-evolved with shared and beneficial services of insects. Various flowers lured certain varieties of insets to flit and transfer fertilizing pollen from one plant of the same species to another.

Unique co-evolved patterns, colors, shapes, nectars and scents of each plant species were observed to be adapted to their counterpart insects' senses. Darwin realized that bees were attracted to blue and yellow flowers while ignoring red ones because they are red-blind.

Other flowers exhibit ultraviolet markings to guide bees that can see beyond the violet range to their nectars.

Butterflies, with good red vision, fertilize red flowers but may ignore the blue and violet ones. Flowers pollinated by night-flying moths tend to lack color but to exude their scents at night. And flowers pollinated by flies, which live on decaying matter, may mimic the (to us) foul smells of putrid flesh.

Darwin was fascinated about ways certain plants were able to ensnare and devour specific insects with moving parts, as well as to see where they were growing. He conducted detailed studies of these remarkable animal-like capabilities in numerous plant research facilities located on his private property.

Some unknown mechanism enabled a Venus flytrap—a member of the sundew family—to clasp its leaves together on an insect to imprison it the moment its trigger-like hairs were touched. These reactions were so fast that Darwin wondered whether electricity could be involved, something perhaps analogous to an animal nerve impulse.

Darwin's colleague Burdon Sanderson was later able to show that electric current was indeed generated by the leaves and could also stimulate them to close. As Darwin recounted in *Insectivorous Plants*:

*The current is disturbed in the same manner as
takes place during the contraction of the muscle
of an animal.*[63]

Darwin also studied climbing plants including twining varieties, leaf-climbers and plants that climbed with the use of tendrils, to investigate how they seemed to know which direction was most advantageous to grow. Regarding the latter, he wrote to his close friend, British botanist and fellow explorer J.D. Hooker, "I believe, Sir, that tendrils can see." [64]

Referring to twisting movements as circumnutation, Darwin observed that when plants grow towards the light, they do not just thrust upwards, they corkscrew. As noted in his last botanical book published in 1880, *The Power of Movement in Plants,* this light-seeking characteristic is evidenced in the earliest evolved plants, including cycads, ferns and seaweeds.

Darwin imagined that the tips of tendrils, a photosensitive region, served as a sort of eye at the tips of seedling leaves. To test this idea he devised little caps, darkened with India ink, to cover them so that they could no longer respond, and they didn't. Darwin then concluded that when light fell on the leaf tip it stimulated the release of some sort of messenger which reached the motor part of the seedling.

Similarly, Darwin found that the primary roots (or radicle), which must negotiate all sorts of obstacles, were also extremely sensitive to contact, gravity, pressure, moisture and chemical gradients. He wrote:

*There is no structure in plants more wonderful,
as far as its functions are concerned, than the tip
of the radicle...It is hardly an exaggeration to say
that the tip of the radicle...acts like the brain of
one of the lower animals...receiving impressions
from the sense-organs, and directing the several*

movements.

Fifty years later it was discovered that plants transmit hormones such as auxins which, in plants, play many of the same roles that nervous systems do in animals.[65]

Darwin always stressed the continuity of life: how all living things are descended from a common ancestor and how we are in this sense all related to each other. This being the case, humans are related not only to apes and other animals but to plants too. As a matter of fact, animals and plants share 70 percent of their DNA.

Charles Darwin's theory of evolution through natural selection based directly upon observed common and differentiated features of geographically isolated living species continues to be validated by fossil discoveries and DNA genetic sequencing revelations. According to English evolutionary biologist at New College, Oxford Richard Dawkins, such information demonstrates beyond all sane doubt the evolutionary basis for our history.[66]

Although certainly not the first time, or likely the last, Darwin's new theories rankled many in the scientific establishment. Philip Henry Gosse, a great naturalist, was so torn by the debate over evolution by natural selection that he wrote a book *Omphalos* claiming that fossil records didn't really correspond to any creatures that ever lived. Instead, they were merely put in the rocks by the Creator to rebuke our curiosity.[67]

New discoveries don't necessarily support what we may ideologically or intuitively wish to believe. For example, Albert Einstein had a violent distaste for the implications of quantum mechanics theory, even though he was one of the very first to demonstrate its counterintuitive processes.

As neurologist and writer Oliver Sacks writes:

Still, in Einstein's time, it was increasingly clear

that the old mechanical, Newtonian worldview was insufficient to explain various phenomena—among them the photoelectric effect, Brownian motion, and the change of mechanics near the speed of light—and had to collapse and leave a rather frightening intellectual vacuum before a radically new concept could be born.[68]

Einstein also took pains to say that a new theory doesn't necessarily invalidate or supersede the old, but rather, allows us to regain our old concepts from a higher level. He wrote:

To use a comparison, we could say that creating a new theory is not like destroying an old barn and erecting a skyscraper in its place. It is rather like climbing a mountain, gaining new and wider views, discovering unexpected connections between our starting point and its rich environment. But the point from which we started out still exists and can be seen, although it appears smaller and forms a tiny part of our broad view gained by the mastery of the obstacles on our adventurous way up.

Or as Darwin's friend, T.H. Huxley, would have put it: "the slaying of a beautiful hypothesis by an ugly fact."

Darwin said: "No one can be a good observer unless he was an active theorizer." Accordingly, his theory of evolution comports well with empirical proofs evidenced by reproducibility and agreement with observations which provide a platform for gaining deeper knowledge. Although this theory status does not signify absolute truth, unlike hypotheses, it has come close enough to support other substantial scientific discoveries.

An example is the discovery of the structure of DNA by James Watson and Frances Crick which gave evolution emphatic validity. Key to molecular biology, this development has enabled detailed understanding of the human genome, genetic diseases and disorders and most appropriate therapeutic treatments.[69]

Genes are specific sequences of chemicals called nucleotides which account for physical variations among people such as skin, eye and hair color. About three billion nucleotides reside within each of twenty-three chromosomes contained in every human sperm cell or egg. Whereas chromosomes among people are largely the same, only one in about each 1,000 nucleotides show a difference.

So far, only one gene is known to have survived unchanged in the biosphere since 1 billion years ago—histone$_4$ codes for DNA molecule housing. All others have and will continue to change. And yet, because of that great evolutionary engine of natural selection, every species became unique, just as each individual is also.

Evolution's implication is that all life ascended from a common single-celled ancestor through a bottom-up, self-organization, coevolution with other life forms, sometimes by trial and error and sometimes interceded by accidents. This whatever-works-best concept along with the projected billions of years required for higher forms to emerge was difficult to imagine in brains conditioned to short-term thinking during Darwin's time.

Truth be told, much like quantum mechanics, the full implications still remain difficult or impossible for human minds to contemplate.

As Oliver Sacks reflected about his own childhood:

Magnolias, my mother explained, were among the most ancient of flowering plants and had appeared nearly a hundred million years ago, at a time when 'moder' insects like bees had not yet evolved, so they had to rely on a more ancient insect, a beetle, for pollination. Bees and butterflies, flowers with colors and scents, were not preordained, waiting in the wings—and they might never have appeared. They would develop together, in infinitesimal stages, over millions of years. The idea of a world without bees or butterflies, without scent or color, affected me with a sense of awe.

Sacks continued:

The world that presented itself to us became a transparent surface, through which one could see the whole history of life. The idea that it could have worked out differently, that dinosaurs might still be roaming the Earth or that human beings might never have evolved, was a dizzying one. It made life seem all the more precious and a wonderful, ongoing adventure ('a glorious accident,' as Stephen Jay Gould called it)—not fixed or predetermined but always susceptible to change and new experience.

Geologic timescales show enormous time required for events that led to Earth's present biotic distribution. The period before the Cambrian, (the Precambrian), lasted about three billion years—approximately 80 percent of Earth's existence. It took that long before there was sufficient atmospheric oxygen to enable the evolution of complex life.

Overall, life has existed for close to 4 billion years, or about

Larry Bell

87 percent of Earth's existence. Blue-green algae fossils about 3.5 billion years old have been found to occur in rocks in Australia and Africa. Chemical traces and carbon isotopes discovered in ancient rocks in Greenland suggest that primitive life including photosynthesizers may have originated even earlier, at least 3.8 billion years ago.[70, 71]

Based upon this incomprehensible time scale, modern humans developed very recently—roughly 160,000 years ago. Our time existence relative to Earth's age is infinitesimal, amounting to only 0.000375 percent of our planet's existence, or about 0.0000107 percent of the Universe's existence.[72]

Our ancestral Hominid line has been around longer, about 2.5 million years. Even then, we're youngsters compared to the age of the dinosaurs which existed some 160 million years.

Earth's life occurred in many exotic places. Archaeans, alae and microbial flora are abundant in the Dead Sea which is ten times saltier than the ocean.

Life exists in sea ice, deep below the surface in basalt and in acidic hot springs. Bacterium Colwellia live in briny liquids bubbles at -20 degrees C in sea ice and can survive in liquid water three molecules thick on mineral surfaces at -196 degrees Celsius.

Tune-like worms and clams aggregate around high temp and pressure and acidic hydrothermal vents on the ocean floor. Halobacteria use light energy without the benefit of chlorophyll—the only non-chlorophyll system known to do so.[73]

Disasters and Do-Overs

Shown by molecular biology, all life evolved from a common ancestor—from a single-celled organism called LUCA (Last Universal Common Ancestor).

DNA reflects linkage from life's beginning to today's diversity.

Life existed for nearly three billion years in a largely oxygen-free environment. Complex life awaited widespread distribution of photosynthetic life.

Ancient life called stromatolites emerged around 3.5 billion years ago. Colonial forms, stromatolites, photosynthesize with the energy of sunlight and emit oxygen. By 1.9 billion years ago, Earth's atmosphere contained some oxygen enabling the first emergence of complex life.

It took nearly another three billion years for enough of that atmospheric oxygen to accumulate that enabled the first complex multi-cellular reproducing life forms to recently develop little more than 530 million years ago.

In the Cambrian explosion a menagerie of fossils became preserved in the Burgess shale of British Columbia. These fossils reflect appearance of many new animal phyla.

Even soft parts became preserved in full for some animals by special conditions in the shale formation. Life's explosion led to twenty more phyla in the Cambrian than in today's biota.[74]

Mass extinctions at intervals of millions of years since that time have had major evolutionary influences, eradicating some life forms and providing advantages for the reemergence of new species to take places of those lost. Modern humans are very much included among the big beneficiaries.

There are likely to be many individual events, including asteroid impacts, climate changes, reformation of continents and toxic gas releases from volcanic eruptions. Some factors remain highly speculative with many unanswered questions.

Scientists have discovered at least five mass extinctions when anywhere between 50 percent and 75 percent of life were lost. These "Big Five" are summarized from oldest to most recent in the following.[75]

Ordovician-Silurian Extinction—439 Million Years Ago

An Ordovician event which began around 439 million years ago wiped out an estimated 86 percent of all species on Earth. Key theories attribute the extinctions to two major causes: glaciation and falling sea levels.

It is believed that Earth was covered at the time with such a vast quantity of plants which removed so much carbon dioxide from the air that temperatures fell drastically.

Falling sea levels which may have resulted from the formation of the Appalachian mountain range may have taken a toll on the majority of early marine animals.

Late Devonian Extinction—349-364 Million Years Ago

Although not known whether this extinction occurred as a single brief event or one spread over hundreds of thousands of years, it may have terminated as many as 75 percent of all species.

Some studies attribute the cause to giant land plants with deep roots that released nutrients into oceans that resulted in algal blooms which depleted the waters of oxygen, thus, animal life.

Another contributing factor may have been volcanic ash which cooled the Earth's temperature, killing off spiders and scorpion-type creatures that had emerged on land by this time.

An ancient amphibian cousin, the elpistostegalians, which had also made it to land became extinct, and vertebrates didn't appear again on land until about 10 million years later. Land animals became common again only after the early Carboniferous period 345 million years ago. This second land colonization correlates with oxygen levels close to today's levels.

Had the late Devonian extinction not occurred, these ichthyostegalians from which we all evolved might not have happened, and we humans would not exist today.

Permian-Triassic Extinction—251 Million Years Ago

Considered the worst in history, this extinction terminated an estimated 96 percent of all species including ancient corals. Today's corals are an entirely different group.

Often referred to as The Great Dying, it is believed to have been caused by an enormous Siberian volcanic eruption that filled the air with carbon dioxide which fed bacteria that emitted huge amounts of methane.

According to theory, the Earth warmed, oceans became acidic, and life descended from only the four percent of the surviving species. Following the event, complex marine life rapidly developed, and snails, urchins and crabs emerged as new species.

Triassic-Jurassic Extinction—119-214 Million Years Ago

This event may have killed off as many as 80 percent of species that had survived the previous Permian-Triassic extinction, opening a path forward for the evolution of dinosaurs that later existed for around 135 million years.

A key cause is broadly attributed to an asteroid impact.

Although small mammals outnumbered dinosaurs during the beginning of this era, by the end, dinosaurs' ancestors (archosaurs) reigned supreme.

Cretaceous-Paleogene Extinction—65 Million Years Ago

The most famous of the Big Five extinctions brought on the extinction of dinosaurs along with an estimated 76 percent of all other life on Earth including many mammals. Attributed to a combination of a ten-kilometer asteroid impact in the Yucatan and volcanic activity, it allowed for the evolution of new mammal species and fish such as sharks at sea.

An asteroid of similar size is expected to impact Earth on average every 50 million to 100 million years. Huge impact craters exist in Woodleigh, Australia and Manicouagan, Quebec, Canada. The latter impact may have occurred 200 to 250 million years ago, perhaps also aiding the Permian-Triassic extinction.

During the Cretaceous period, many shallow seas flooded parts of continents leading to extensive marine sediment formation. Because fossil preservation occurs in sedimentary rocks, a better record of Cretaceous life exists than for other mass extinctions.

Oxygenated Pathways to Modern Life Forms

Reptiles appeared in the late Triassic period about 300 million years ago during what is referred to as the Mesozoic Era, a time of high oxygen levels. Some 230 million years ago they developed a new respiratory system with small, rigid sac-like appendages called a septate added to old lungs to handle lowering oxygen levels.

This change led to the appearance of dinosaurs because this more efficient oxygen-handling system enabled them to survive the mass extinction at the end of the Triassic. The more advanced lungs also supported dinosaur evolution into birds. Lizards, on the other hand, didn't acquire this evolutionary advantage, which is why they cannot thrive in high altitude thin air.

Birds, the surviving descendants of dinosaurs, still have lungs of dinosaur structure. These air sacs are filled with air when birds breathe, where it is stored a short time before passing to their lungs for the next inhalation. The highly efficient exchange between air and blood enables birds to even to extract far more oxygen than animals of comparable size. This explains why geese can fly over the Himalayas at altitudes which are lethal to humans.[76]

Many paleontologists believe that periodic ups and downs in oxygen levels have affected much of the previous and existing extreme variation life forms.

About 542 million years ago atmospheric oxygen was lower than today and fluctuated for the next 100 million years. Afterward oxygen rose steadily until 400 years ago. It was some 25 percent at the beginning of the Devonian period. Then a steep decline occurred, followed by a rise peaking near the end of the Carboniferous. Then, oxygen levels are believed to have been near 30 percent. This was again followed by a nadir at about 12 percent oxygen at the Triassic end—causing a mass extinction—followed by an irregular rise to today's 21 percent level.

Thanks largely to an oxygen-rich atmosphere, the Cambrian period which commenced about 542 million years ago witnessed the rapid evolution of new life forms. At least a dozen or more new animal phyla, each with very different body plans, emerged over the time space of a million years or less—a geological eye blink.

The once rather peaceful pre-Cambrian seas were transformed into a liquid jungle of newly-mobile hunters and hunted. And while some animals (such as sponges) lost their nerve cells and regressed to a vegetative life, others, especially predators, evolved increasingly sophisticated sense organs, memories and vegetative minds. (The latter concept of "mind" can be envisioned by considering a human body lacking most brain activity but kept alive by maintaining unconscious brain region functioning.)

In his last book, *The Formation of Vegetable Mould, Through the Action of Worms,* Darwin wrote that worms, which can distinguish between light and dark and modulate their responses to threats suggests "the presence of a mind of some kind."

Darwin noted that worms generally stay underground safe from predators during daylight hours, and while they have no ears, are very sensitive to vibrations conducted through the earth such as footsteps of approaching animals. All of these sensations, Darwin concluded, are transmitted to collections of nerve cells (he called them the cerebral ganglia) in the worm's head.

Following an 1859 discovery by Louis Agassiz that the jellyfish Biugainvillea had a substantial nervous system containing about a thousand nerve cells, George John Romanes—Darwin's young friend and student—demonstrated in 1883 all that the jellyfish employed both autonomous, local network-dependent mechanisms and centrally-coordinated activities through the circular brain-like organ that ran along margins of the bell.

Romanes wrote:

> *[N]erve tissue is invariably present in all species whose zoological position is not below that of the Hydroza. The lowest animals in which it has hitherto been directed are the Medusae, or jellyfishes, and from them upwards its occurrence is, as I have said, invariable.*
>
> *Wherever it does occur its fundamental structure is very much the same, so that whether we meet with nerve tissue in Jellyfish, an oyster, an insect, a bird, or a man, we have no difficulty in recognizing its structural units as*

everywhere more or less similar.[77]

Although perhaps not very smart, jellyfish do seem to have minds of their own. For example, they can change direction and depth, and many have a "fishing" behavior that involves turning upside down for a minute, spreading their tentacles like a net, and then righting themselves, which they do by virtue of eight gravity-sensing balancing organs. Box jellyfish (Cubomedusae) have fully developed image-forming eyes, not so different from our own.

At the same time that Romanes was vivisecting jellyfish and starfish, a passionate young Darwinist named Sigmond Freud was studying cells of vertebrates and invertebrates in the Vienna laboratory of psychologist Ernst Brucke. Freud was particularly interested in comparing a very primitive vertebrate (Petromyzon, a lamprey) with those of an invertebrate (a crayfish).

Freud produced meticulous, beautiful illustrations showing all nerve cells were found to be basically similar...including with those of human beings. The nerve cell body and its processes— dendrites and axons—constitute the common basic building blocks, serving as each nervous system's signaling units.

Although neurons may differ in shape and size, they are essentially the same from the most primitive animal to the most advanced. Only their number and organization significantly differ. Whereas we have a hundred billion nerve cells, while a jellyfish has a thousand, their status as cells capable of rapid and repetitive firing is essentially the same.[78]

The crucial role of synapses—the junctions between neurons—to modulate individual organism behaviors was clarified only at the close of the nineteenth century by the great Spanish anatomist Santiago Ramon y Cajal. English neurophysiologist Charles Sherrington coined the word synapse, demonstrating that synapses could be excitatory or inhibitory in function.

Within a few years of Darwin's death, it had been discovered that even single-celled organisms like the protozoa could exhibit a range of adaptive responses. A tiny, stalked, trumpet-shaped unicellular Stentor organism employs a repertoire of at least five different responses to being touched before finally detaching itself to find a new site if these basic responses are ineffective.

If touched again, the Stentor will skip the intermediate steps and immediately take off for another site. In this sense, it has become sensitized to remember an unpleasant experience and learn from it, although the memory lasts for only a few minutes.[79]

In the 1960s, Eric Kandel embarked on a study of the cellular basis of memory and learning based upon examination of a giant sea snail, the Aplysia. The mollusk was selected because it has a relatively few (20,000 or so) neurons which are distributed in ten or so ganglia of about 2,000 neurons apiece. Aplysia has particularly large neurons—some that are visible to the naked eye—connected with one another in fixed anatomical circuits.

Kandel wrote:

> *I appreciated that all animals have some form of mental life that reflects the architecture of their nervous system.*[80]

The ancient Aplysia exhibits a protective reflective reflex that, when threatened, withdraws its exposed gill to safety. By recording and sometimes stimulating nerve cells and synapses in the abdominal ganglion that governs these responses, Kendel was able to show that it exhibited both short- and long-term memory. He observed that the short-term memory, as involved in habituation and sensitization, depended on functional changes in synapses. Longer-term memory, which might sometimes last several months, evidenced structural changes in the synapses. In neither case was there any change in the actual circuits.[81]

Whereby Aplysia has only about 20,000 neurons distributed in ganglia throughout its body, an insect, despite tiny size, may have up to a million nerve cells which enable extraordinary cognitive feats. As previously discussed, bees are expert in recognizing different colors, smells and geometrical shapes presented in a laboratory setting, as well as systematic transformations of these. In addition, they not only recognize the appropriate patterns and smells and colors, but can also explore and remember their locations, and communicate these coordinates to fellow bees.

Laboratory observations have demonstrated that members of a highly social paper wasp species can even recognize individual faces of others. Such face-learning cognitive skill have been broadly associated only with mammals.

As for exhibiting any conclusive definition true consciousness, the very term begs clear interpretation. For example, Charles Darwin noted in *The Voyage of the Beagle* that an octopus in a tidal pool seemed to interact with him in a watchfully curious and even playful manner.

As also observed by others, domesticated cephalopods often seem to evoke a similar sense of empathy and emotional proximity with humans that we associate with consciousness of felines and dogs.[82]

Nature has employed at least two very different ways of making a brain—indeed, there are almost as many ways as there are phyla in the animal kingdom. Yet as Oliver Sacks reminds us, mind, to varying degrees, has arisen or is embodied in all of these, despite the profound biological gulf that separates them from one another, and us from them.

Modern humans have inherited a large neurological advantage. Our brains containing upwards of a hundred trillion neurons, each with up to ten thousand synapses, afford practically infinite capacities for conscious and unconscious manipulations.

Unfathomably, each of these neuronal signaling groups remain in constant communication with each other, weaving, many times a second in continuously changing but always meaningful patterns.[83]

Our astonishingly numerous synapses enable us to remember things, albeit, not as accurately as we might often imagine that we do. Freud was among the earliest to postulate that an important physiological prerequisite for memory was a system of contact barriers between neurons—his so-called psi system. This was a decade before Sherrington gave synapses their name.

No one was more sensitive than Freud to the reconstructive potential of memory—to the fact that memories are continually worked over and revised through recategorization and re-transcription. Freud wrote in a letter to Wilhelm Fliess in 1896:

> *As you know, I am working on the assumption that our psychic mechanism has come into being by a process of stratification, the material present in the form of memory traces being subject from time to time to a rearrangement in accordance with fresh circumstances—a transcription...Memory is present not once but several times over...the successive representations representing the psychic achievement of successive epochs of life...I explain the peculiarities of the psychoneuroses by supposing that this translation has not taken place in the case of the same material.*

British psychologist Frederic Charles Bartlett, who conducted experimental studies in the 1930s, concluded that memory imaginatively constructs and reconstructs endlessly. He wrote:

Remembering is not the re-excitement of innumerable fixed, lifeless and fragmentary traces. It is an imaginative reconstruction, or construction, built out of the relation of our attitude towards a whole active mass of organized past experience, and to a little outstanding detail which commonly appears in image or in language form. It is thus hardly ever exact, even in the most rudimentary cases of recapitulation, and is not all important that it should be so.

Neurologist Oliver Sacks adds that in filling in a memory blind spot, the brain constructs a plausible hypothesis or pattern or scene. In the absence of outside confirmation, there is no easy way of distinguishing a genuine memory or inspiration, felt as such, from those that have been borrowed or suggested between a "historical truth" and "narrative truth."

Sacks writes:

Indifference to source allows us to assimilate what we read, what we are told, what others say and think and write and paint, as intensely as if that were primary experiences. It allows us to see and hear with other eyes and ears, to enter into other minds, to assimilate the art and science and religion of the whole culture, to enter into and contribute to the common mind, the general commonwealth of knowledge. Memory arises not only from experience but from the intercourse of many minds.

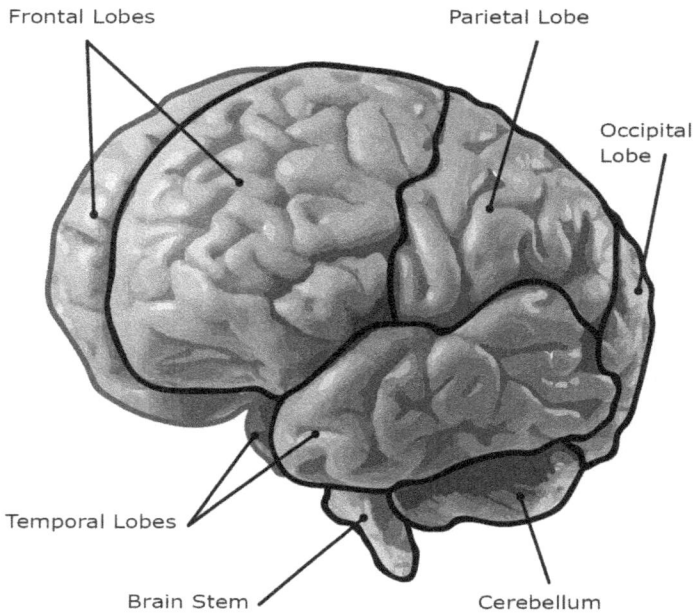

BRAIN ANATOMY

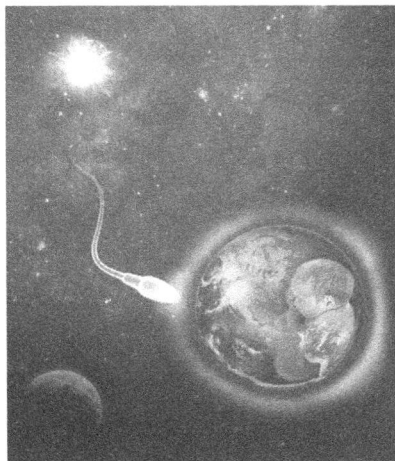

Part Three: The Birth of Consciousness

OF COURSE WE wouldn't be aware of remembering anything, or even contemplate about life, our own very much included, were it not for possessing this marvelous quality of human consciousness. But what, exactly, is that? What evolutionary requirements must be met for it to exist? And in what ways are the human variety and capabilities of consciousness different—superior—to other living creatures?

So, let's briefly fast forward now in this evolutionary life narrative to consider consciousness from a contemporary Homo sapiens perspective. After all, and whatever it is, without its capacities I wouldn't be writing this book at my present moment and you wouldn't be reading this sentence right now at your separate personal moment.

Be warned, however, that the what's, how's and why's of human conscious—all consciousness—remain highly speculative. Human brains, curiosities, experiments and imaginations have concocted many theories about what makes us tick and tickle; what makes each of us our special and unique selves. Truth be known, although we have learned much about some of the neuro autonomic architecture of the conscious brain, we still don't really know why we became so incredibly fortunate to have one.

What it Means to be Conscious

Consciousness is a relative term which has been applied at some level or other to most animal species, including certain invertebrates such as the octopus. Although not very well understood, neuroscientists generally tend to loosely associate these different levels of consciousness with the number of the brain's active neurons. Species with only a few neurons are typically considered to be unconscious automatons.[84]

The late psychologist and lecturer at Princeton University Julian Jaynes argued that even if we knew the connections of every tickling thread of every single axon and dendrite in every species that ever existed, together with all the neurotransmitters and how they varied in billions of synapses of every brain that ever existed, we could still never—not ever—from a knowledge of the brain alone know if that brain contained a consciousness like our own.

In the introduction of his book *The Origin of Consciousness in the Breakdown of the Bicameral Mind,* Jaynes poetically describes human consciousness as a "world of unseen visions and heard silences, this insubstantial country of the mind!" [85]

Jaynes expands this state of being, asking:

> *What ineffable essences, these touchless rememberings and unshowable reveries! And the privacy of it all! A secret theater of speechless monologue and prevenient counsel, an invisible mansion of all moods, musings, and mysteries, an infinite resort of disappointments and discoveries. A whole kingdom where each of*

us reigns reclusively alone, questioning what we will, commanding what we can. A hidden hermitage where we may study out the troubled book of what we have done and yet may do. An introcosm that is more myself than anything I can find in a mirror. This consciousness that is myself of selves, that is everything, and yet nothing at all—what is it?

And where did it come from? And why?

Despite centuries of experiments and ponderings, fundamental questions keep returning without answers. Jaynes continues:

[Consciousness] is the difference that will not go away, the difference between what others see of us and our sense of our inner selves and the deep feelings that sustain it. The difference between the you-and-me of the shared behavioral world and the unlocatable location of things thought about. Our reflections and dreams, and the imaginary conversations we have with others, in which never-to-be-known-to-anyone we excuse, defend, proclaim our hopes and regrets, our futures and our pasts, all this thick fabric of fancy is so absolutely different from handable, standable, kickable reality with its trees, grass, tables, oceans, hands, stars—even brains!

Jaynes emphasizes that humankind has pondered the fundamental essence of consciousness almost since consciousness began, each age describing the term according to its particular theme and concerns. He then outlined eight influential concepts:

- Consciousness as a [Fundamental] Property of Matter:

Jaynes characterized this concept—which is sometimes characterized as neo-realism—as being particularly popular during the first quarter of the last century. It tends to view consciousness primarily in mathematical terms much as a phenomenon of particle physics.

- Consciousness as a Property of Protoplasm:

This view asserts that rather than existing in all matter, consciousness is a fundamental property of all living things whereby the smallest one-celled animals advanced in continuous evolution through coelenterates, the protochordates, fish, amphibians, reptiles and mammals—ultimately to man. This idea gained momentum based largely upon observations of lower organisms by noted scientists including Charles Darwin and E.B. Titchener during the early twentieth century.

Julian Jaynes was highly skeptical about imputing consciousness to protozoa, stimulation—responsive behaviors which reside entirely in physical chemistry, not in introspective psychology. He also regarded it as a big stretch to empathetically identify human consciousness with the agony of a struggling worm on a fish hook. If the worm felt pain as we do, the writhing tail end of a worm severed from the "brain" end would have been disconnected from conscious sensory actuation. The wriggling results from a mechanical release phenomenon as motor nerves in the tail end fire in volleys when cut off from normal inhibition by the cephalic ganglion.

- Consciousness as Learning:

This concept holds that to make consciousness coexistent with protoplasm leads to a third presumption that consciousness began not with matter, nor at the beginning of animal life, but at some specific time later after life had sufficiently evolved. It followed that if an animal could modify its behavior on the basis of experience, it then must be having an experience that it was conscious of.

Jaynes argued that the error in this thinking is to confuse consciousness with an actual space inhabited by elements called sensations and ideas, and the association of these elements because they are like each other, or because they have been made by the external world to occur together, is indeed what learning is all about. This causes learning and consciousness to become muddled together with that vaguest of terms, experience. Instead, the evolution of the origin of learning and the origin of consciousness are two entirely separate matters.

- Consciousness as a Metaphysical Imposition:

Unlike the previous theories that assume that consciousness evolved biologically by simple natural selection, another position denies that such an assumption is possible.

Alfred Russel Wallace, who along with Darwin co-discovered the theory of natural selection, insisted that man's conscious faculties "could not possibly have been developed by means of the same laws which would have determined the progressive development of the organic world in general, and also of man's physical organism."

Wallace believed that some metaphysical force had directed evolution at three different points: the beginning of life, the beginning of consciousness and the beginning of civilized culture. He spent the latter part of his life searching in vain among séances of spiritualists for evidence to support this proposition.[86]

- The Helpless Spectator Theory:

English philosopher Shadworth Hodgson is among the prominent nineteenth century theorists who believed that what we do is completely controlled by the wiring diagram of the brain and its reflexes to external stimuli, as accordingly, consciousness is metaphorically nothing more than the heat given off by those wires.

Or as Jaynes describes this concept:

> *Consciousness is the melody that floats from the harp and cannot pluck its strings, the foam struck raging from the river that cannot change its course, the shadow that loyally walks step for step beside the pedestrian, but is quite unable to influence his journey.*

- Emergent Evolution:

This simple idea goes back to John Stuart Mill and G.H. Lewes as popularized in Lloyd Morgan's 1923 book *Emergent Evolution* which posits that all properties of matter emerged from some unspecified forerunner: those complex chemical compounds emerged from conjunctions of simpler components; and properties of distinctive living emerged from conjunctions of complex molecules; and consciousness emerged from living things.

New emergent properties are in each case effectively related to the systems from which they emerge, whereby new relations emergent at each higher-level guide and sustain the course of events distinctive at that level. Accordingly, consciousness emerges as something genuinely new at a critical stage of evolutionary advance.

Julian Jaynes then asks if consciousness emerged in evolution, when did this happen? In what species? What kind of nervous system is necessary? He pointed out:

> *What is wrong about emergent evolution is not the doctrine, but the release back into old comfortable ways of thinking about consciousness and behavior. The license that it gives to broad and vacuous generalities.*

- Behaviorism:

Jaynes traced the philosophical roots of behaviorism back to so-called Epicureans of the eighteenth century and before in attempts to generalize tropisms from plants, to animals, to man—to movements called Objectivism. At first it began very similar to the helpless spectator theory which held that consciousness just was not important in animals. It later asserted that consciousness is really nothing at all!

As Jaynes describes this, following toppled idealism after WWI, behaviorism "allowed a new generation to sweep aside with one impatient gesture all the worn-out complexities of the problem of consciousness and its origin."

Jaynes continued:

> *Off the printed page, behaviorism was only a refusal to talk about consciousness. Nobody really believed he was not conscious…In essence, behaviorism was a method, not a theory that it tried to be. And as a method, it exorcised old ghosts. It gave psychology a thorough house cleaning. And now the closets have been swept out and cupboards washed and aired, and we are ready to examine the problem again.*

- Consciousness as the Reticular Activating System:

This approach to understanding consciousness generally assumes that modern techniques now available to study complex brain anatomy and processes will enable science to find where it is that consciousness resides. Jaynes reflects that this search dates back to Descartes' identification of the brain's pineal body as the seat of consciousness, a theory roundly refuted by physiologists of his time…and that search is still on.

The present possible neural substrate site nominee for

consciousness is a tangle of tiny internuncial neurons in the brain stem called the reticular formation. Surprisingly, from an evolutionary standpoint, this turns out to be one of the oldest—perhaps the very oldest—parts of our nervous system.

Often referred to as the reticular activating system, this formation extends from the top of the spinal cord through the brainstem on up into the thalamus and hypothalamus, connecting with collaterals from sensory and motor nerves. It also has direct lines of command to a half a dozen major areas of the brain's cortex.

The reticular activating system's main function is to sensitize or awaken selected nervous circuits and desensitize others. It is the place where general anesthesia produces its effect by deactivating its neurons. Cutting it produces permanent sleep and coma, whereas stimulating it through an implanted electrode in most regions awakens a sleeping animal.

Although interesting, what little is presently known about the evolution of the reticular function doesn't appear to answer fundamental questions regarding the essence of what constitutes consciousness. Complex psychological phenomena do not readily translate into neuro-anatomical and chemical constructs.

Jaynes describes some features of consciousness:

- Spatialization:

Spatialization is a characteristic of all conscious thought where you first habitually turn your mind-space attention to abstract things that can be *separated out, put aside each other*, and *looked at*—as could never happen physically or in actuality. You then make the metaphor of the theories as concrete objects, then the metaphor of a temporal succession of such objects as a synchronic array, and thirdly, the metaphor of the characteristics of the objects, all of some degree so that they can be *arranged* in a kind of logical *fit*.

- Excerption:

In consciousness we are never *seeing* anything in its entirety because in actual behavior we can only see or pay attention to part of a thing at any one moment. So, in consciousness we excerpt from a collection of possible attentions to a thing which comprises our knowledge of it.

Thus, if someone asks you to think of a circus, for example, you may have a fleeting moment of slight fuzziness, followed by picturing a trapeze artist or clown in the center ring. We are never conscious of things in their true nature, only of the excerpts we make of them.

Excerption is distinct from memory. Jaynes posited:

> *An excerpt of a thing is in consciousness the representative of the thing or event to which memories adhere, and by which we can retrieve memories.*

- The Analog "I":

Jaynes describes a most important feature of consciousness is the metaphor we have for ourselves, the analog *I* which can "move about" vicariously in our imagination, *doing* things we are not actually doing. Here, we imagine *ourselves* doing this or that, and thus make decisions on the basis of imagined outcomes.

For illustration, Jaynes asks us to imagine that we are out walking, and two roads diverge in a wood, and we know that one of them comes back to our destination after a much more circuitous route, we can *traverse* that longer route with our analog *I* to see if its vistas and ponds are worth the time it will take.

- The Metaphor "Me:"

The metaphor *I* is also a metaphor *me,* where as we imagine ourselves strolling down the longer path we catch *glimpses* of *ourselves* from the imagines vistas…we can perhaps step back a bit and see ourselves kneeling down for a drink of water at a particular brook. We *put* ourselves in the picture.

- Narratization:

In consciousness we are constantly seeing our vicarial selves as the main figures in the stories of our lives, although it is not so obvious that we are doing this. We selectively perceive new situations as part of this ongoing story, and perceptions that don't fit into it tend to go unnoticed, or at least unremembered. The assigning of causes to our behavior or saying why we did a particular thing is all part of this narration, although such reasons may be true or false, neutral or ideal.

Jaynes proposed that consciousness is always ready to explain anything we happen to find ourselves doing, just as a thief narrates his act as due to poverty, the poet his as due to beauty, and the scientist his as due to truth, purpose and cause.

- Conciliation:

This aspect of consciousness which springs from simple recognition is modeled upon a behavioral process common to most mammals whereby a slightly ambiguous object is made to conform to a previously learned schema. This brings things together as conscious objects just as narration brings things together as a story.

In conciliation we are making excerpts or narrations compatible with each other. If you are asked to think of a mountain meadow and a tower at the same time, you may automatically conciliate them by seeing the tower rising from a meadow. However, if you are asked to think of a mountain meadow and an ocean at the same time, conciliation tends not to occur and you are likely to think of one, and then the other.

My neuropsychologist friend and well-known bestselling author David Eagleman reflects that while scientists often debate the detailed definition of human consciousness, it's easy enough to pin down what we're usually talking about with the help of a simple comparison: when you're awake you have consciousness, and when you're in deep sleep you don't. That distinction gives us an inroad for a simple question: what is the difference in brain activity between these two states? [87]

Neuroscientist Susan Greenfield describes dreaming and consciousness as being similar. During wakefulness, the prefrontal cortex becomes highly active, but in dreams the cortex is under-active and the limbic system dominates. The dream state is similar to schizophrenia—strong emotion without logic. Without prefrontal cortex restraint, dreaming is phantasmagoric. Generally irrational, dreaming sometimes promotes new ideas for creative problem solution. [88]

The Human Advantage

So, if dreaming, which occurs when we are in a sense unconscious and in an autonomous mode, what is the purpose of wakeful consciousness?

David Eagleman suggests that from an evolutionary point of view, the purpose of consciousness is to enable rapid switching between situational survival requirements and-goal setting priorities in response to novel and unexpected challenges. In the animal kingdom, most animals do certain things very well (say, prying seeds from the inside of a pine cone), while only a few species (such as humans) have the flexibility to dynamically develop new mental software programs.

Eagleman reminds us that we spend most of our awake moments operating pretty much on autopilot. When we walk down a city street, we seem to automatically know what things are without having to work out the details. Our brains make assumptions about what we're seeing based upon internal models built up over years of experience walking other city streets.

Every experience we've had contributes to the internal models in each of our brains.

Dr. Eagleman observes that the first thing we learn from studying our own circuitry is a simple lesson: most of what we do and think and feel is not under our conscious control:

> *The vast jungles of neurons operate their own programs. The conscious you—the 'I' that flickers to life when you wake up in the morning—is the smallest bit of what's transpiring in your brain. Although we are dependent on the functioning of the brain for our inner lives, it runs its own show. Most of its operations are above the security clearance of the conscious mind. The 'I' simply has no right of entry.*

Consciousness gets involved when the unexpected happens, when we need to work out what to do next. Although the brain tries to tick along as long as possible on autopilot, it's not always possible in a world that throws curveballs. In this sense, Eagleman characterizes our consciousness as being like a tiny stowaway on a transatlantic steamship, taking credit for the journey without acknowledging the massive engineering underfoot.[89]

Fundamentally, the brain is tuned to detect unexpected outcomes—and this sensitivity is at the heart of animals' ability to adapt and learn. It's no surprise, then, that the brain architecture involved in learning from experience is consistent across species, from honeybees to humans. This suggests that brains discovered the basic principles of learning from reward long ago.

Eagleman equates the wakeful consciousness mind to the role of the CEO of a corporation. As long as the zombie mental subroutines are running smoothly, the CEO can sleep. It is only when something goes wrong (say, all the departments suddenly find their business models have catastrophically failed), that the CEO is called. He writes:

> *Think about when your conscious awareness comes online: in those situations where events in the world violate your expectations. When everything is going according to the needs and skills of your zombie systems, you are not consciously aware of most of what's in front of you; when suddenly they cannot handle the task, you become consciously aware of the problem. The CEO scrambles around, looking for fast solutions, dialing up everyone who can address the problem best.*[90]

As CEO, our consciousness takes charge of long-term planning while most of the day-to-day operations are run by all the parts of the brain to which this executive has little or no access. Eagleman asks us to imagine a CEO who has inherited a giant blue-chip company. Although she or he has some influence, they are coming into a situation that has already been evolving for a long time before they consciously arrived. Their new job is to define a vision insofar as the technology of the company is able to support these policies. This is what consciousness does: it sets goals, and the rest of the system learns how to meet them.

In meeting these challenges, our executive consciousness also enables us to do a form of simulated time travel. Dr. Eagleman invites us to consider this scenario:

> *I have a bit of free time and I'm trying to decide what to do. I need to get groceries, but I also know I need to get to a coffee shop and work on a grant for my lab, because a deadline is coming*

*up. I also want to spend time with my son in the
park. How do I arbitrate this menu of options?*[91]

It would be easy, of course, if we could directly compare these
experiences by living each one, and then rewinding time, and
finally choosing our path based upon which outcome was the
best. So, whether wisely or not, our executive CEO
consciousness takes responsibility to make such decisions for us
based upon prior experience outcomes and consequences.

So, time travel is something the human brain does
relentlessly. When faced with a decision, our brains simulate
different outcomes to generate a mock-up of what our future
might be. This enables us to mentally disconnect from the
present moment and voyage to a world that doesn't exist yet.

How all of this works is a grand mystery. Research based
upon models of animal neuro-anatomy and behavior tell only a
small part of an unknowably large story. Studies, for example, of
how a sea slug withdraws from touch, how a mouse responds to
rewards and how an owl localizes sounds in the dark ultimately
only reveal little more than blueprints of circuitry that respond
to particular inputs with appropriate outputs.

If our brains were composed only of these patterns and
circuits, would it feel like anything to be alive and conscious?
Eagleman asks, wouldn't it feel like nothing—to be like a
zombie?

And what about those other less conscious animals—how
do they feel? Although science has no meaningful way to make
measurements to answer that question, Eagleman offers two
intuitions:

> *First, consciousness is probably not an all-or-
> nothing quality, but comes in degrees. Second, I
> suggest that an animal's degree of consciousness
> will parallel its intellectual flexibility. The more*

subroutines an animal possesses, the more it will require a CEO to lead the organization. The CEO keeps the subroutines unified; it is the warden of the zombies. To put this another way, a small corporation does not require a CEO who earns three million dollars a year, but a large corporation does. The only difference is the number of workers the CEO has to keep track of, allocate among, and set goals for.[92]

Eagleman proposes that a useful consciousness index is an animal's capacity to successfully mediate conflicting zombie systems. The more an animal looks like a jumble of hardwired input-output subroutines, the less it gives evidence of consciousness; the more it can coordinate, delay gratification and learn new programs, the more conscious it may be.

He cites, the herring gull as a gullible example. If you put a red egg in its nest, it goes berserk. The red color triggers aggression in the bird, while the shape of the egg triggers brooding behavior. As a result, it tries to simultaneously attack the egg and incubate it.[93]

Writer Oliver Sacks posited that a fundamental prerequisite for consciousness is an ability to fuse discrete instantaneous, momentary, visual frame experiences into a continuously flowing mobile state of awareness. He believes that such dynamic consciousness probably first arose in reptiles a quarter of a billion years ago.

Sacks argued that it seemed probable that no such stream of consciousness exists in amphibians. A frog, for example, shows no active attention and no visual following of events. Rather than scanning its surroundings to look for prey, it relies upon what appears to be a purely automatic ability to recognize insect-like objects that enters its visual field and to dart out its tongue in response.

Executive consciousness functions build upon more complex learned behaviors in more advanced animals. Nobel Prize winner Gerald Edelman, in his book *Wider Than the Sky: The Phenomenal Gift of Consciousness*, asks us to imagine a jungle animal with primary consciousness level, one affording a coherent, unified world scene:

> It hears a low growling noise, and at the same time the wind shifts and the light begins to wane. It quickly runs away, to a safer location. A physicist might not be able to detect any necessary causal relation among these events. But to an animal with primary consciousness, just such a set of simultaneous events might have accompanies a previous experience, which included the appearance of a tiger.
>
> Consciousness allowed integration of the present scene with the animal's past history of conscious experience, and that integration has survival value whether the tiger is present or not.[94]

None of this would be possible without complex sensory systems that enable animals to receive information from the outside world in ways that can be processed to evoke appropriate responses As David Eagleman points out:

> Just look across the animal kingdom, and you'll find a mind-boggling variety of peripheral sensors in use by animal brains. Snakes have heat sensors. The glass knifefish has electrosensors for interpreting changes in the local electrical field. Cows and birds have magnetite, with which they can orient themselves to the Earth's

magnetic field. Animals can see ultraviolet;
elephants can hear at very long distances, while
dogs experience a richly scented reality.[95]

Our human perception of reality regarding the world around us is interpreted by our senses as well. In a larger sense, the truly real world is a materially empty, colorless and soundless place.

While we tend to think that stones are hard, snow is cold and grass is green, laws of physics inform us otherwise.

Everything we perceive as hard is made of tiny bits of energetic stuff that has no solidity at all; the temperature we sense is but specific electromagnetic frequencies and wavelengths detected by neurons and calibrated by our brains to inform us about our surroundings; and color perception is an interpretative phenomena as well. Yellow light, for example, describes transversal electromagnetic wavelengths in the neighborhood of 590 nanometers. Those images we see in front of us are really illusory constructs which are assembled behind our eyes in our brains. The sensory impressions of taste and smell depend upon specialized chemical receptors that differentiate between presence of different molecules or ions. Hearing, a mechanical process, relates to how our brains interpret various vibrations within the range of our audio detectors caused when something moves air around.

Eagleman observes that these different sensory input-response devices demonstrate an enormous variety of ways the crucible of natural selection has enabled genetic codes to channel data from surrounding outside worlds into internal worlds to reveal many different slices of reality. He concludes from this that there is nothing remarkably special or fundamental about the sensors we're used to:

They're just what we inherited from a complex
history of evolutionary constraints.

The Matter of Mind over Matter

A decade ago, neuroscientists Francis Crick and Christof Koch asked, "Why does our brain consist simply of a series of special zombie systems?" In other words, why are we conscious of anything at all? Why aren't we simply a collection of these automated circuits routines that solve problems? [96]

The mathematician-philosopher Rene Descartes assumed that there was something far more to consciousness than this... an immaterial soul which exists separately from the brain. He even speculated that this spiritual sensory input enters through the brain's pineal gland.

Descartes' theory of dualism suggested that there are two realms of existence. One of these is a physical realm comprised of the environment and things around us—a realm of matter and energy. This realm can be scientifically researched because it operates in a prescribed and mechanical way. The other realm is "transcendent" to the physical environment, hence, cannot be measured.

Dualism became an influential concept during Descartes' time because it allowed scientists to conduct research without fear of being charged with heresies by religious groups. Correspondingly, the theory created a problem for psychologists and students of the mind who could no longer view mental behaviors in purely mechanical terms.

Sigmund Freud, who came along later, suspected that diverse varieties of human behaviors were really explicable only in terms of unseen mental processes, the machinery running things behind the scenes.

Could thinking really be equated with the processing done by the nervous system?

Could the mind really be like a machine?

By carefully examining his patients, Freud observed that

there was often nothing obvious in their conscious minds driving their behavior, and so, given the new machinelike view of the brain, he concluded that there must be underlying causes that were hidden from access. In this new perspective, the mind was not simply equal to the conscious part we familiarly live with; rather it was like an iceberg, the majority of its mass hidden from sight.

Because Freud lived many decades before modern brain technologies, his best research option was to gather data from "outside" of the system: by talking to patients and trying to infer their brain states from mental states. He then hypothesized that these states existed as products of hidden neural mechanisms... machinery to which the subject had no direct access.

David Eagleman emphasizes that neither Descartes nor Freud had access to wander modern neurology wards to witness ways that various brains' physical and electrical neuroanatomy can be studied to reveal how tiny factors affect dramatic changes in personality and mental states.

Some kinds of brain damage, for example, make people depressed. Other changes make them manic. Others adjust a person's religiosity, sense of humor, or appetite for gambling. Others make a person indecisive, delusional or aggressive. Hence, Eagleman argues great difficulty in validating entirely separate scientific frameworks for mental and physical states.

Referring to what he characterizes as a mind-body problem, Eagleman urges us to recognize that understanding human consciousness requires a need to think not in terms of the pieces and parts of the brain, but instead, to think in terms of how these components interconnect.[97]

The typical brain has about 86 billion neurons, each making about 10 thousand connections which collaborate in a very specific manner that is unique to each person. Our experiences, our memories, all the stuff that makes us who we are is represented by the unique pattern of the quadrillion connections between our brain cells.

Eagleman offers an analogy for how this all comes together into higher consciousness by broadly comparing the process to swarm intelligence expressed by an ant colony. Although the colony as a collective whole accomplishes extraordinary feats, each ant individually behaves simplistically. It just follows simple rules.

The queen doesn't give commanding orders; she doesn't coordinate the behavior from on high. Instead, each ant reacts to local chemical signals from other ants, larvae, intruders, food, waste, or leaves. Each ant is a modest, autonomous unit whose reactions depend only on its local environment and the genetically encoded rules for its variety of ant.

When enough ants come together, a superorganism emerges with collective properties that are more sophisticated than its basic parts. This phenomenon, known as emergence, is what happens when simple units interact in the right ways so that something larger arises.

And so it goes with the brain. A neuron is simply a specialized cell, just like other cells in your body, but with some specializations that allow it to grow processes and propagate electrical signals. Like an ant, an individual brain cell just runs its local program its whole life, carrying electrical signals along its membrane, spitting out neurotransmitters when the time comes for it and being spat upon by the neurotransmitters of other cells.

Eagleman concludes:

> *That's it. It lives in darkness. Each neuron spends its life embedded in a network of other cells, simply responding to signals. It doesn't know if it's involved in moving your eyes to read Shakespeare, or moving your hands to play Beethoven. It doesn't know about you. Although your goals, intentions, and abilities are*

completely dependent on the existence of these little neurons, they live on a smaller scale, with no awareness of the thing they have come together to build.[98]

Special Consciousness of Being Me and You

So with all those robotically programmed ant-like neurons, why aren't we all just wondering around like mindless zombies? Even more, what makes each of a uniquely self-consciously-identifying me?

Eighteenth century German philosopher Emanuel Kant referred to this self-awareness as an unknowable *transcendental apperception of the ego*—seeing this self not as something perceived by the senses, but rather as something spiritually eternal which transcends us. In his 1781 book *A Critique of Pure Reason,* he maintained that the mind relies upon a priori forms, abilities and ideas supplied by divine acts which were always within us, and from which everything else flows.[99]

Princeton neurosciences Professor Michael Graziano also contemplates relationships between our physical brains and transcendental thoughts. In his book *Consciousness and the Social Brain,* he broadly defines this personal consciousness of self as the window through which we understand:

> *The essence of self-awareness…the spark that makes us us…something lovely that apparently is buried inside us that makes us aware of ourselves and the world.*[100]

In asking the question "who am I?", David Eagleman reportedly answered himself, saying:

> *When I think about who I am, there's one aspect*

above all else that can't be ignored: I am a sentient being. I experience my existence. I feel like I'm here. Looking out on the world through these eyes, perceiving this Technicolor show from my own center stage. Let's call this feeling consciousness or awareness.[101]

Australian-born philosopher Ludwig Wittgenstein observed, just as the eye, which is the source of the visual field but not in the visual field, cannot see itself, so it is with the *I* which is the source of our consciousness.

This condition strikes me as being somewhat analogous to a Geico insurance television commercial where we see a gecko lizard character walking through a stone tunnel towards an opening. When he approaches it, we witness him looking out searching for Mount Rushmore from the entrance connection in George Washington's eye which he was unaware of from the inside.[102]

But still, how did each of us come to experience that conscious "me-ish" awareness that sets us apart from everyone else? For example, why is it that we can feel things that touch us, but not things that touch someone else?

American psychologist and humanist philosopher Carl Rogers described such a self-concept as:

...the organized conceptual gestalt composed of the characteristics of 'I' or 'me' and the perceptions of the relationships of the 'I' or 'me' to others and to various aspects of life, together with the values attached to these perceptions. It is a gestalt which is available to awareness though not necessarily in awareness. It is a fluid and changing gestalt, a process, but at any given moment is a specific entity.[103]

So yes, each of us somehow experience our own gestalt, which in turn is somehow part, and perhaps in some ways also independent of an unknowably incomprehensible large and timeless gestalt. Some may refer to this as God, others as nature and maybe still others as a quasi-mechanical or metaphysical phenomenon.

Nevertheless, we don't necessarily have to understand how things work—or why—in order to marvel at the fact that they exist. Descartes, in his *Meditations on First Philosophy*, appreciated that while he could doubt that he had a body (it could just be a dream or illusion created by an evil demon), he could not doubt that he had a mind…a *thinking thing* which carried the essence of himself which doubts, believes, hopes and contemplates.

Plato's allegory *Phaedo* likens our lack of understanding of the objects and phenomena by which we perceive our world to emerging from a dark cave into sunlight where only vague shadows of what lies beyond that prison are cast dimly upon the wall. Those shadow forms of perception are both non-physical and non-mental, existing nowhere in time, space, mind or matter.

Einstein referred to wonders of the Universe as an un-openable pocket watch. And although he spent much of his career arguing against quantum theory which he himself had contributed much to develop, he nevertheless acknowledged its advantages in explaining subatomic phenomena. Most importantly, it worked.

V.S. Ramachandra, author of *The Tell-Tale Brain: A Neuroscientist's Quest for What makes Us Human*, marvels, as we all should, about what makes our high level of human consciousness possible. He asks:

> *How can a three-pound mass of jelly that you can hold in the palm imagine angels,*

contemplate the meaning of infinity, and even question its own place in the cosmos? Especially awe inspiring is the fact that any single brain, including yours, is made up of atoms that were forged in the hearts of countless, far-flung stars billions of years ago. These particles drifted for eons and light years until gravity and change brought them together here, now. These atoms form a conglomerate—your brain—that can not only ponder the very stars that gave it birth but can also think about its own ability to think and wonder about its own ability to wonder. With arrival of humans, it has been said, the Universe has suddenly become conscious of itself. This, truly, is the greatest mystery of all.

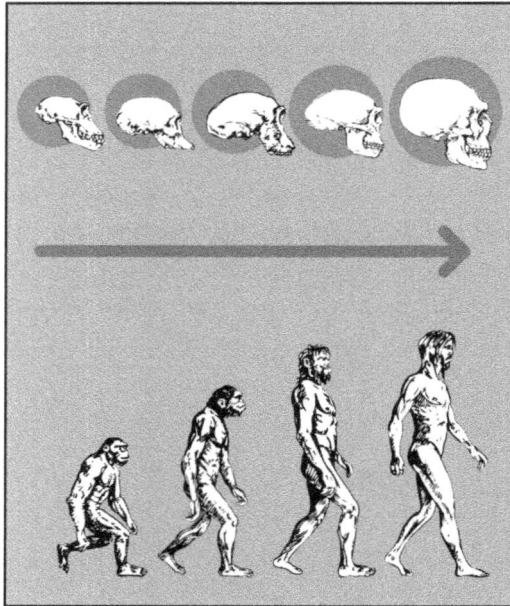

EVOLUTION OF THE HUMAN BRAIN

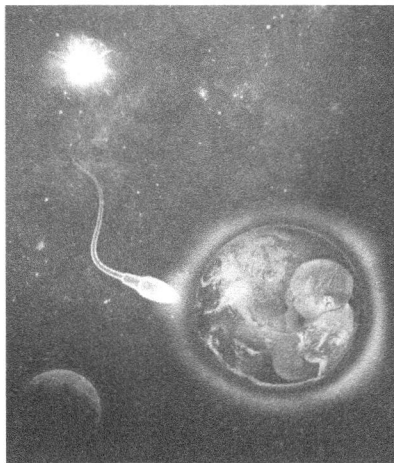

Part Four: Our Ancestors and Hominid Cousins

IN HER BOOK *The Human Advantage*, Suzana Herculano-Houzel observed:

> As far as I know, and as presumptuous as it may
> seem, it is a fact that we are the only species to
> study itself and others to generate knowledge
> that transcends what is observed firsthand; to
> tamper with itself, fixing imperfections with the
> likes of glasses, implants, and surgery and thus
> changing the odds of natural selection; and to
> modify its environment so extensively (for
> better or for worse), extending its habitat to
> improbable locations.
>
> We are the only species to use tools to make
> other tools and technologies that extend the
> range of problems it can tackle; to further its
> abilities by seeking harder and harder problems

> to solve; and to invent ways to register
> knowledge and to instruct later generations that
> go beyond teaching by direct demonstration.
> Even though all this may be achieved through no
> particular cognitive ability exclusive to our
> species, we certainly take these abilities to a
> level of complexity and flexibility that is rivaled
> by none.[104]

How did this marvelous cognitive engine that only emerged a few millennium ago develop to drive the evolution of technology, language and complex social behavior? Carl Sagan attributed its origin to a common ancestral proto-reptile:

> Somewhere in the steaming jungles of the
> Carboniferous period there emerged an
> organism that for the first time in history of the
> world had more information in its brains than in
> its genes. It was an early reptile which, were we
> to come upon it in these sophisticated times, we
> would probably not describe as exceptionally
> intelligent.
>
> But its brain was a symbolic turning point in
> the history of life. The two subsequent bursts of
> brain evolution, accompanying the emergence of
> mammals and the advent of manlike primates,
> were still the important advances in the
> evolution of intelligence.[105]

Referring to experiments, such as those conducted on squirrel monkeys by Paul McLean, chief of the Laboratory of Brain Evolution and Behavior of the National Institute of Mental Health, Sagan subscribed to an evolutionary brain model analogous to three interconnected biological computers. This

"Tribune" theory postulated that each of the three modern brain components corresponded to a separate evolutionary advancement step. The most ancient of these, a reptilian complex—or R-complex—connected immediately to the spinal cord, contains just the basic neural machinery for reproduction and self-preservation. These functions included regulation of the heart, blood circulation and respiration.

According to McLean, a second more advanced, limbic system evolved to surround the R-complex about one hundred and fifty million years ago. We share this development with other mammals, although not "in its full elaboration with the reptiles."

The final brain component stage to emerge in the triune theory is the neocortex which exists in higher mammals and primates. Humans have the largest and most fully developed neocortex for their size, followed by dolphins and whales.

Suzana Herculano-Houzel strongly takes issue with what she describes as an intuitive, but incorrect Tribune theory that mammalian brains evolved from earlier reptilian predecessors.

She writes:

> The Tribune brain is a fantasy, however. As more fossils of sauropsids (the proper name for dinosaurs) were uncovered, some of them feathered. It became clear that modern lizards, crocodiles, and birds are close cousins, all now considered reptiles (birds included), whereas mammals arose separately, and very early on, from a sister group at the very beginnings of amniote life. Mammals were therefore never reptiles or birds back in evolutionary time; the brain of mammals is at least as old as the brain of birds and other reptiles, if not older—it just evolved along a different evolutionary path.

Dr. Herculano-Houzel adds that if mammals did not descend from reptile-like beings, they could not have a brain that was built by adding layers on top of that reptilian-like brain:

> *Comparing the mammalian brain to that of the reptile brain and presuming one to have layered new structures on top of the other is just as preposterous as looking at two living cousins and expecting one of them to have given birth to the other.*[106]

Whereas elementary life has existed on this planet for close to four billion years, the first hominids emerged in East Africa on a rapid evolutionary growth path from an earlier genus apes called *Australopithecines* (meaning "Southern Ape") only as recently as about 2.5 million years ago.

By about 55 million years ago, during the Eocene period, there was a great proliferation of primates, both arboreal and ground-dwelling, and the evolution of a line of descent that eventually led to man. Based upon endocranial fossil evidence, a prosimian called *Tetonius* exhibited tiny nubs where brain frontal lobes later evolved.

The first fossil evidence of a brain of even vaguely human aspects dates back to eighteen million years to the Miocene period, when an anthropoid ape called *Proconsul* appeared. Proconsul was probably ancestral to the present great apes and possibly also to modern humans as well.

DNA evidence indicates gorillas severed from our lineage 7 to 9 million years ago, while chimpanzees and bipedal apes went their separate ways 5 to 7 million years ago. The gap between bipedal apes and humanoids is the notorious missing link.[107]

In July 2002, a French paleoanthropologist named Michael Brunet and his team found a complete, seven-million-year-old Australopithecines cranium with canine teeth in Chad's Djurab

Desert in Central Africa which they named *Sahelanthropus tchadensis*. The site was 1,500 miles west of the East Africa Rift Valley and South Africa, where previous searches for early hominids had been concentrated.

Of special interest, the partial skull showed a modification in the direction of humans with its lower face projecting less than apes. Brunet concluded that the newly discovered species might be a close relative to our common ancestor with chimpanzee— the earliest member of human lineage. Characteristic of discoveries, Brunet's contention did not meet the science community without dissent.[108]

Although it is debatable by some whether Australopithecines were directly ancestral to modern humans or merely close collateral relatives, casts from their fossil skulls reveal that like us, they possessed large brains for their body weight. It is certain, however, that a great abundance of apelike animals existed by about five million years ago.

Some Australopithecines clearly not of the genus Homo (not human) were still incompletely bipedal with brain masses only about a third of the size of the average adult modern human. Were we to meet an Australopithecine today, we would perhaps be struck by an almost total absence of forehead.

A French-American team found an ape-like-faced Australopithecus fossil with small canine teeth nicknamed Lucy (LUCA, our Last Universal Common Ancestor) that lived in Ethiopia and Kenya roughly between three and 3.7 million years ago. Although bipedal, the species had long arms to enable adept tree climbing and chipped stones for rudimentary tools.

The Australopithecines *robustus* possessed impressive nut-cracker teeth and a remarkable evolutionary stability. Its endocranial volume varied very little from specimen to specimen over millions of years of time. This species was taller and heavier than the *Gracile* Australopithecines variant that walked on two feet and had brain volumes of about 500 cubic centimeters, some

100 cubic centimeters larger than brains of the modern chimpanzee.

Australopithecines *gracile* appear to be a considerably older species, with much more variance in endocranial volumes over time than their robust cousins. Judging again from their teeth, the Gracile probably ate meat as well as vegetables.

The A. Gracile fossil sites reveal implements made of stone and animal bones, horns and teeth which were painstakingly carved, broken, rubbed and polished to make chipping, flaking, pounding and cutting tools. No such tools have been associated with the A. Robustus species, whose brain to body weight was only about half as large as the A. Gracile.

It is natural to speculate that this large relative difference in brain size may have a connection between having tools or no tools. Paleontologists also theorize that bipedalism may have preceded encephalization, by which our ancestors and cousins walked on two legs before they evolved big brains.[109]

A remarkable aspect of the archaeological record concerning tools is that as soon as they appear at all they appear in enormous abundance. This suggests the existence of ancient stonecraft education programs which passed on the skills between tribes from generation to generation.

Less than three million years ago, the glacial-interglacial times of the Pleistocene Epoch emerged, whereby climate changes caused both chimpanzee and australopithecine populations to become downsized and forced into isolated areas in order to sustain traditional ways of life. This isolation may have enabled a new variant of australopithecine to emerge about 2.5 million years ago—the indisputable beginning of the Homo lineage having bigger brains with cortical folds.[110]

African forests shrank, influenced by a long, cool and dry period. And while some species adapted, others were forced to develop new food sources. By three million years ago, some Australopithecines were in a transition from a vegetarian to an omnivorous diet. It is theorized that extra energy from eating meat allowed brain growth, influencing the first undeniable hominid on the scene with larger cranial volumes than the East African A. Gracile.

One of them, which Richard Leakey of the National Museums of Kenya called *Homo habilis*, had a brain volume of about 700 cubic centimeters. Archeological evidence shows that like A. Gracile, H. Habilis also made a variety of tools. This observation supports the theory first advanced by Charles Darwin that tool-making is both the cause and effect of walking on two legs, which frees the hands.[111]

In addition, arrangements of stones at H. Habilis sites indicate that they may have constructed dwellings as well dating back more than two million years—long before the Pleistocene Ice Ages when humans sought refuge in caves.

Were we to encounter a H. Habilis today dressed in contemporary garb—we would probably give them little notice other than due to their relatively small stature. They had a high forehead like modern humans, suggesting a significant development of the neocortical areas in brain frontal and temporal lobes.

Anthropologist Ralph L. Holloway of Columbia University believes that a region of the brain known as Broca's area, one of several centers required for speech, can also be detected in H. Habilis fossil evidence. This finding suggests that the development of language, tools and culture may have occurred simultaneously.

H. Habilis, which is widely credited as the first true human, emerged during the same epoch as A. Robustus. Being larger both in body and brain weight than either the Australopithecines, the species possessed a ratio of brain to body weight about the same as that of the Gracile.

Since H. Habilis and A. Robustus emerged at the same time, it is very unlikely that one was the ancestor of the other. The A.

Gracile were also contemporaries of H. Habilis but much more ancient. It is therefore possible—yet uncertain—that both H. Habilis, with a promising evolutionary future, and A. Robostus, an evolutionary dead end, arose from an earlier Gracile, Australopithecines *africanus*, who survived long enough to be their contemporary.

H. Habilis inhabited vast African savannahs filled with an enormous variety of predators and prey. The first modern horse appeared at about that same time. Although the horses survived, H. Habilis died out about 1.6 million years ago.[112]

In 1984, Kamoya Kimeu, a member of a team led by Richard Leakey, discovered a nearly complete 1.6-million-year-old skeleton of a young boy at Nariokotome near Lake Turkana in Kenya they named the Turkana Boy whose features showed considerable modern human identity. Leakey termed the species *Homo erectus* (Upright Man).

Possessing endocranial brain volumes similar to ours, various other excavations sites reveal that H. Erectus had developed a sophisticated toolkit with lozenge-shaped stones, cleavers and implements.

Chinese H. Erectus specimen locations which are clearly associated with the remains of cave campfires indicate that members of this group termed the Peking Man domesticated fire more than a half-million years ago.

Emerging Out of Africa

About two million years ago, some of these archaic humans left their homeland to journey through and settle vast areas of North Africa, Europe and Asia. The new human pioneers divided into a variety of different species.

Humans in Europe and western Asia evolved into *Homo neanderthalensis* (Man from the Neander Valley) popularly referred to as *Neanderthals*. The more eastern regions of Asia became populated by H. Erectus who survived there for close to 2 million years, making it the most durable human species ever.

Early humans adapted to a variety of environmental conditions. *Homo solensis* (Man from the Solo Valley), who was well suited to life in the tropics, inhabited the Indonesia island of Java.[113]

Those on another Indonesian island—the small island of Flores—underwent a process of dwarfing. These *Homo floresiensis* who had first arrived when the sea level was exceptionally low, making the island easily accessible from the mainland, became trapped after the seas again rose. Since the island was poor in resources, smaller members of their group who required less food survived better than those of larger body stature.

Although the resulting dwarf population reached a maximum height of only 3.5 feet and weighed no more than fifty-five pounds, they were nevertheless able to produce stone tools. They also managed to hunt a dwarf species of the island's elephants.[114]

Our own species, *Homo sapiens* (immodestly termed Wise Man) only very recently became part of the natural scene. Although it's not certain exactly where and when animals that came to be popularly known as Sapiens first evolved from some earlier type of humans, most scientists agree that by 150,000 years ago, East Africa was populated by people that looked just like us.[115]

Scientists also agree that about 70,000 years ago, Sapiens from East Africa spread into the Arabian Peninsula, and from there they quickly overran the entire Eurasian landmass. By this time, most of Eurasia was already settled by other humans who soon disappeared.

What happened to the others? As noted in Noah Harari's popular book *Sapiens: A Brief History of Mankind*, there are two conflicting theories: one that they interbred and merged with

Sapiens, and another that the Sapiens exterminated competitors.

A replacement theory presents a story of attraction, sex and mingling which presumes that the species were genetically close enough to produce fertile offspring. Here, a species is defined by a group that can produce fertile offspring by crosses within—but not outside—itself. For example, the mating of different breeds of dogs yields puppies which, when grown, will be reproductively competent dogs. On the other hand, crosses between species— even species as similar as donkeys and horses—produce infertile offspring (in this case, mules).

Accordingly, donkeys and horses are categorized as separate species. And although mating of more widely separated species— for example lions and tigers—have been reported to (very rarely) produce fertile offspring, this indicates that sometimes the strict species definition can become fuzzy.

Technically, all human beings are considered to be members of the same genus (Homo), while some including H. Habilis, a recently discovered *Homo denisova*, Erectus and Neanderthals have been classified as different species than Sapiens, some questions arise whether the first of these categories could have interbred with Sapiens to produce our ancestors.

In the case of H. Denisova, Erectus and Neanderthals, however, DNA evidence indicates that there can be little doubt. It seems that about 50,000 years ago, Sapiens, Denisovans, Erectus and Neanderthals were at a borderline point…almost, but not quite entirely separate species.

Tests show that between one and four percent of the unique human DNA of modern populations in the Middle East and Europe is Neanderthal. DNA extracted from a fossilized H. Denisova finger bone found in a Siberia cave showed that up to 6 percent of the unique human DNA of modern Melanesians and Aboriginal Australians is Denisovan.[116]

As human immigrants spread around the world, they bred with other human populations to ultimately produce us. For example, the more inventively versatile Sapiens that reached the Middle East and Europe encountered and bred with Neanderthals who were more muscular and bodily adapted to cold. From this standpoint, today's Eurasians aren't entirely "true Sapiens," but rather, are a mixture. Similarly, DNA samples indicate that when Sapiens reached East Asia they bred with the local Erectus, so that Chinese and Koreans are a mixture of Sapiens and Erectus.

All encounters between these different groups, however, were clearly not of a romantic nature, breeding only fatal results for Sapiens competitors. As a likely consequence, H. Denisova disappeared slightly less than 50,000 years ago, followed by Neanderthals about 30,000 years ago.

Sapiens had already populated East Africa 150,000 years ago, and by about 70,000 years ago, began to overrun the rest of planet Earth and drive the other human species to extinction.

In the first recorded encounter between Sapiens and Neanderthals, the Neanderthals won. About 100,000 years ago, some Sapiens groups migrated north to the Levant, which was Neanderthal territory, but failed to secure firm footing. Although not known why, the Sapiens retreated, leaving Neanderthals as masters of the Middle East.[117]

Beginning about 70,000 years ago, Sapiens bands that left Africa for a second time drove the Neanderthals and all other human species not only from the Middle East, but altogether from the face of the Earth. They accomplished this applying the same strategic methods they had perfected in hunting non-human prey.

Whereas Neanderthals usually hunted alone or in small groups. Sapiens had developed techniques that relied on cooperation between many dozens of individuals, and perhaps even between different bands.

One particularly effective method was to surround an entire herd of animals, such as horses, then chase them into a narrow gorge, where it was easy to slaughter them en masse. The bands then collaborated with one another to harvest and share tons of meat, fat and animal skin. Archaeologists have even discovered sites where fences and obstacles were erected to create artificial traps and slaughtering grounds.[118]

Although we may presume that Neanderthals were not pleased to see their traditional hunting grounds turned into Sapiens-controlled slaughterhouses, faced with more advanced collaborative battle strategies, the Neanderthals were not much better off than other wild game. Whereas a Neanderthal would probably have beaten a Sapiens in a one-on-one brawl, they wouldn't stand a chance when their individuals or groups were attacked by hundreds.

Neanderthals were unable to effectively coordinate defenses involving large groups, nor could they adapt their social behavior to innovative Sapiens challenges. Even if the Sapiens lost a first round, they could quickly invent new stratagems that would enable them to win the next time.

Such cooperation gave Sapiens a crucial edge over other human species. It is believed that relations with neighboring bands were sometimes tight enough that together they constituted a single tribe, sharing a common language and common norms of behavior.

How We Sapiens Got Our Big Heads

Throughout most of hominid history, our ancestral brain sizes, proportional to overall bodies, were approximately the same as other apes. Skull sizes of contemporary chimpanzees and Australopithecus, an extinct hominid that lived between 3.0 and 2.9 million years ago, averaged around 450 milliliters. By about 1.8 million years ago, hominid brains were averaging around 600 milliliters.

Paleolithic records reveal that as Homo erectus emerged with a 750 cubic centimeter brain, the neocortex—often characterized as the rational brain—began expanding particularly significantly. Found only in animals, this six-layered sheet-like structure in the roof of the forebrain (and particularly the outsized prefrontal cortex), enables us to hold and manipulate sophisticated concepts. Although we can't match a jaguar in a foot race, we can easily outrun all other animals in agile mental simulations.

Our modern neocortex mushroomed to its current size less than one million years ago, a remarkably short time considering our future-looking, tool-wielding, symbol-juggling human family broke off from the great apes in Africa at least six million years earlier. This human brain growth evolution was attended by neural system wiring changes that augmented processing power for functional tasks. There was a particularly significant development advancement in the Broca's area, the part of the frontal lobe that is correlated with language.[119]

Brain size growth also came with a necessity to enlarge the female pelvic girdle structure to accommodate birthing of larger-headed babies.

Present adult men and women have braincases twice larger than the volume of recorded for Homo habilis. The incomplete closure of the modern human skull at birth, the fontanelle, is very likely an imperfect accommodation for this recent brain evolution.

A big mental capacity jump during the Middle Pleistocene period between 800,000 and 200,000 years ago doubled average hominid brain sizes. Anthropologist Rick Potts, who directs the Human Origins Program at the Smithsonian Natural History Museum, has attributed this exploded size and complexity to survival requirements imposed by intense climate change adaptation pressures involving thinking patterns, behaviors and even social strategies. Habituations that had previously worked well for those living in jungle environs became ineffective in the woodland fringe environments, on the grassland savannah or on the coasts.[120]

Unfamiliar environmental and unstable situations during that period may have forced Homo erectus to slow down in order to analyze and strategically plan new solutions where purely habit-based strategies no longer worked. Basic survival required a more powerful information processor with better memory, reasoning abilities and social skills needed for rapid adaptation to unfamiliar environmental and resource conditions.

Our successful Sapiens ancestors became adept at dealing with these challenges, while the Neanderthal, our close genetic Eurasian cousins (sharing about 90 percent of the same genes) were not. Although they too used tools, buried their dead, built fires and were bigger and stronger than our ancestors, failures to adapt to climate change ended their existence.

As their world became colder, whereas the Neanderthals retreated into shrinking forests, the more inventive Sapiens expanded their food hunting and gathering territories into growing tundra environments. And whereas the physically more robust physiques of Neanderthals fit well in the context of forest ambush-style hunting short-range, the Sapiens innovated more effective strategies. They hunted in coordinated groups; developed lighter, longer-range projectiles such as spears; and created better-adapted clothing and shelter "technologies" for the colder regions.[121]

Anthropologist Robin Dunbar attributed much influence over rapid expansion in the size of the human neocortex with increased cognitive demands of cultural learning.

He reasons that as groups became larger and more closely organized during the Middle Pleistocene period between 800,000 and 200,000 years ago, it became more cognitively demanding to keep track of many complex social relationships.

This imposed requirements for increased mental processing power to build better memories, representational abilities and even language.[122, 123]

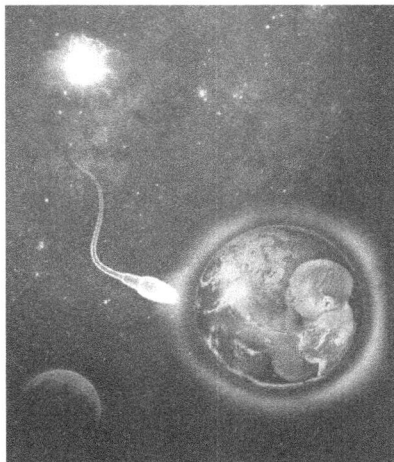

Part Five: The Invention of Culture

REMARKABLE THREE-FOLD HOMINID brain growth over 2.5 million years opened rapid social evolution. Evolving language capacities and growing populations promoted swarm intelligence and tribal cooperation, with super-organisms greatly exceeding individual capabilities.

About 50,000 years ago, greater Sapiens brain capacities enabled a creative explosion that continues today. The Neanderthals' extinction resulted from a failure to apply comparable cultural flexibility enabling them to move beyond experience-based strategies and tried-and-true customs. By 30,000 years ago, Sapiens had displaced Neanderthals, along with all other hominid species.

Neurosurgeon Frank Vertosick characterizes human intelligence as an emergent property from large groups of cooperative societies...an ability to store past experiences for use in future problem-solving.[124]

Daniel Goleman agrees, observing that "new thinking holds that our sociability has been the primary survival strategy of primate species, including our own." [125]

Sociability is a primary human survival strategy that engages the cooperative interactions of many minds. It enables advanced intelligence necessary to learn and teach new skills, to outwit predators and to compete with adversaries. It affords a mechanism of using abilities of innumerable brains much as if it were one collective brain.

Present day human intellect can be viewed as both a product and servant of our social life. Our improvising imagination—our early intellect—gave us the behavioral and mental scaffolding to organize and manage our experiences. This cultural capacity began to occur long before human imagination invented language and word concepts.

Although Sapiens gained a variety of versatile, inventive advantages, other early humans developed primary components of what might legitimately be characterized as cultures as well. For example, archeologists have discovered bones of Neanderthals with severe physical handicaps, suggesting that they were cared for by relatives.

Having said this, no living creature has ever surpassed the Sapiens' ability to adapt to immigrate and innovatively adapt to such a huge variety of radically different habitats so quickly. [126]

Life in the Cognitive Revolution Fast Lane

The immense diversity of Sapiens behavioral adaptations and inventions are primary components of constantly evolving cultures—unstoppable developments referred to as human history.

Our fortunate ancestral cognitive and social inheritance has led to even faster-paced innovations. Imagine that while 11 millennia passed between the Agricultural Revolution and the Industrial Revolution, it only took only 120 to get from the Industrial Revolution to the light bulb. Humans landed on the

Moon 90 years later. Twenty-two years after that came the World Wide Web. A mere nine years later, the human genome was fully sequenced.[127]

Archaeological records reveal few changes in archaic human social patterns or new technologies prior to a Cognitive Revolution which began about 70,000 years ago. Following the emergence of Homo erectus, stone tools which are broadly recognized as a defining feature of our species remained roughly the same over nearly 2 million years.

The emergence of Homo sapiens was a huge cognitive and behavioral game-changer—one requiring no significant new genetic or environmental dependencies.

As noted by Yuval Noah Harari:

> *Consequently, ever since the Cognitive Revolution, Homo sapiens has been able to revise its behavior rapidly in accordance with changing needs. This opened a fast lane of cultural evolution, bypassing the traffic jam of genetic evolution. Speeding down this fast lane, Homo sapiens soon far outstripped all other human and animal species to cooperate.*[128]

Harari characterizes the Cognitive Revolution as the point when history declared its independence from biology. Until then, "the doings of all human species belonged to the realm of biology, or if you so prefer, prehistory." (Harari clarifies that he prefers to avoid the term "prehistory," because it wrongly implies that even before the Cognitive Revolution, humans were in a category of their own.)

Whereas a hike in East Africa two million years ago would likely have revealed what appeared to be a familiar cast of human characters, Harari points out that their behaviors were not really so different than many other animals. Just as archaic humans loved, played, formed close friendships and competed for status and power—so did chimpanzees, baboons and elephants.

The period from about 70,000 years ago to about 30,000 years ago witnessed the invention of boats, oil lamps, bows and arrows and needles (essential for sewing warm clothing). Then within a remarkably short period, about 45,000 years ago, Sapiens had reached Europe, East Asia and had somehow crossed the open sea to Australia—a continent previously untouched by humans.

Sapiens, whose bodies were originally adapted to living in the African savannah, devised ingenious solutions which enabled them to migrate to brutally frigid climates such as northern Siberia which even cold-adapted Neanderthals avoided. They learned to make snowshoes and effective thermal clothing composed of layers of furs and skins, sewn together tightly with the help of needles.

As their thermal clothing and hunting techniques improved, sapiens dared to venture deeper and deeper into the frozen regions. More advanced weapons and sophisticated hunting techniques enabled Sapiens to track and kill mammoths and the other big game of the far north. Every mammoth was a vast quantity of meat (which, given the cold temperatures, could be frozen for later use), tasty fat, warm fur and valuable ivory.

Whereas early Sapiens inhabited existing caves or rock shelters where available, only very recently, especially within the last 20,000 years, were those natural shelters enhanced with interior walls or other simple modifications.

Shelters in open areas were often constructed using a range of framework materials including wooden poles and bones of large animals, such as mammoths. These structures were probably covered with animal hides. Habitat sites sometimes included fire hearths.

Around 14,000 BC, the Sapiens chase took some of them from north-eastern Siberia to Alaska. At first, glaciers blocked the way from Alaska to the rest of America, allowing no more than perhaps a few isolated pioneers to investigate lands further south. However, around 12,000 BC, global warming melted the ice and opened an easier passage.[129]

Making use of the new Bering Straits corridor, our ancestors moved south en masse, spreading over the entire North American continent. And although originally adapted to hunting large game in the Arctic, they soon adjusted to an amazing variety of new climates and ecosystems.

Within merely a millennium or two, Siberian descendants settled the thick forests of what became the eastern United States, the swamps of the Mississippi Delta, the deserts of Mexico and steaming jungles of Central America. Some made their homes in the river world of the Amazon basin, while still others struck roots in Andean mountain valleys and open pampas of Argentina.

Shortly after man entered North America in the Pleistocene period, there were massive and spectacular kills of large game animals, often by driving them over cliffs. This ability to stalk a single wildebeest or to stampede a herd of antelope to their deaths indicates that the hunters shared at least some minimal symbolic verbal language.

While skeletons of H. Erectus suggest the absence of muscle control for respiratory speech recognition, some rudimentary form of vocalization probably existed as much as 2.5 million years ago. Positioned much lower in the throat than for apes, the upright posture of Erectus favored the evolution of a larynx enabling enunciation of consonant and vowel sounds. Also, the upright bipedal breathing posture facilitated specialized muscle development for high-fidelity sounds.

Darwin pointed out that gestural languages cannot usefully be employed while our hands are otherwise occupied, or at night, or when our view of the hands is obstructed. One can then imagine gestural languages being gradually supplemented and then supplanted by verbal languages—which may have originally imitated the sound of the object or action being communicated.

Although any language ability we inherited from H. Erectus, the first Homo species to hunt as a major means of subsistence, would have been very limited, it would still have been superior to that of H. Habilis. In any case, they managed to make sophisticated stone tools, and to extend their range beyond Africa.

Oldwan stone implements, including simple choppers and pounders, continued with little innovation for about a million years between 2.6 and 1.8 million years ago. Subsequent tool-making industries, such as the Acheulean (1.7-0.2 MYA) and Mousterian (30,000-40,000 MYA), were more innovative and diverse.

Production of simple hand axes that were flaked repeatedly to create a biface symmetrical point remained basically the same for over a million years before evolving to create other adaptive tools like spear points, arrowheads and stone knives.

Here, what is typically characterized as the "stone age" might more accurately be described as the "wood age." Since archeological evidence consists mainly of bones and stone tools, artifacts made these ancient hunter-gatherers of more perishable materials such as wood, bamboo and leather would leave fewer fossil traces.

Yet soon after emerging in the last Ice Age more than 30,000 years ago, Cro-Magnon Sapiens in southern France invented elaborate tools, spear throwers, fully barbed harpoons and flint master tools for making hunting weapons. These early humans learned how to secure handles to stone for hand axes with tree resin or bitumen, using a lever principle for increased power and velocity for work and distance killing. They innovated arrows with wooden shafts. They stitched together animal skins with bone needles for caps, shirts, jackets, trousers and moccasins. They built mammoth bone houses, laid stone floors, made animal fat burning lamps and they built fireplaces for heating and cooking.[130]

Visual technologies soon advanced and spread as well. Cave paintings, such as the Chauvet Cave in the Gorges de l'Ardèche, France dating back 30,000 years ago, demonstrate that human minds could convert three-dimensional animals (e.g., a bison or bear) into two-dimensional line representations.[131]

Although Sapiens had relatively very simple cultures compared to now, they were far more advanced than any previous species. As they evolved and expanded, our ancestors brought a trading culture with them. Included were social prestige items such as shells, amber and pigments.

Archaeologists excavating 30,000-year-old Sapiens sites in the European heartland have discovered seashells from the Mediterranean and Atlantic coasts. In all likelihood, these shells got to the continental interior through long-distance exchanges between different Sapiens bands. Neanderthal sites, on the other hand, lack any evidence of such commerce.[132]

Humans are by no means unique among other animals in forming social bonds and relationships; ours are by far most complex. However, we are the only creatures that can truly make any sense out of trading something truly needed—food to eat, or weapon to hunt with, for example—for a bauble of adornment or coin of currency with no practical intrinsic value.

As Yuval Noah Harari notes, a unique Sapiens trait is our ability to create and believe in our own fictions to create imagined realities. All other animals use their communication system to describe reality. While not all fictions are shared by all humans, at least one—money—has become universal. Harari observes that while dollar bills have absolutely no value except in our collective imagination, everybody believes in them.[133]

Such bonds and collaborations between Sapiens provided a crucial edge over other human species. Relations between neighboring tribes eventually became tight enough to merge into single tribes that shared common languages, common myths, common norms and common values. These combined tribes exchanged members, hunted together, traded luxuries, cemented political alliances, celebrated religious festivals and honored sacred rituals.

Humans are the only animal that both anticipates its mortality and contemplates what may follow. Burial ceremonies that include the interment of food and artifacts along with the deceased go back at least to our Neanderthal cousins, suggesting not only a widespread awareness of death, but also an already developed ritual ceremony to sustain the deceased in the afterlife.

Burials were infrequent and very simple prior to 40,000 years ago. They later began to become much more elaborate with the inclusion of valued objects such as tools and body adornments. Red ochre was sprinkled over many of the bodies prior to burial.

In 1955, archaeologists discovered the skeleton of a fifty-year-old man covered with strings containing a total of 3,000 mammoth ivory beads in a 30,000-year-old burial site in Sunghir, Russia. This adornment, together with 25 mammoth ivory wrist bracelets and a hat covered with fox teeth indicated that he was probably an important leader of a hierarchical society.

Archaeologists also discovered another Sunghir tomb containing two skeletons buried head-to-head. One belonged to a boy aged about twelve or thirteen, and the other a girl about nine or ten. The boy who was covered with 5,000 ivory beads wore a hat and a belt containing a total of 250 fox teeth (at least sixty dead foxes). The girl was adorned with 5,250 ivory beads. Both children were surrounded by statuettes and various ivory objects.

One conclusion suggests that both were children of very important tribal figures. An alternate theory is that they may have been ritually sacrificed—possibly as part of ceremonial burial rites of a leader.[134]

Not all encounters between Sapiens bands were friendly.

Some groups violently fought one another over resources and for other unknown reasons.

A 12,000-year-old cemetery in the Sudan evidenced that 24 of 59 skeletons discovered there were found with arrowheads and spear points embedded in them or lying nearby. The skeleton of one woman revealed 12 injuries.

Archaeologists at Ofnet Cave in Bavaria discovered the remains of 38 foragers, mainly women and children, who had been thrown into two burial pits. Half of the skeletons, including those of children and babies, bore clear signs of damage by human weapons such as clubs and knives. Those few belonging to mature males bore the worst marks of violence. It is suspected that an entire forager band had been massacred.[135]

The Evolutionary Art of Language and Innovation

Referring to a Sapiens advantage over Neanderthals he attributed to "cognitive fluidity," author and University of Reading Deputy Vice Chancellor Steven Mithen noted that Sapiens gained a mental capacity which led to an explosion some 50,000 years ago in language, technology and art. Mithen pointed out that whereas Neanderthals lacked the "ability for metaphor and had limited imagination," cognitive fluid thought requires an ability to make connections between modular thinking domains which involve perceptions of social, material and natural worlds.[136]

Stephen Asma cites the development of language as both a cognitive product and enablement of Sapiens' rapid evolutionary success.

He writes:

> *Most evolutionary psychologists claim that the cause of [the Homo sapiens'] cognitive fluidity was the development of language (in the late Pleistocene), because language provides an*

obvious syntactical/grammatical system for manipulating representations.[137]

Asma contemplates:

> *I try, for example, to imagine what it was like to be a conscious being before language (either a Homo erectus man or a contemporary Homo sapiens baby), I run straight into the fact that my mind is already deeply structured by language. It is difficult to peek around the veil of language to see the pre-linguistic operating system at work.*[138]

Eric Kandel literally visualizes an answer to Asma's dilemma. In his book *The Age of Insight*, he writes:

> *Perhaps in human evolution the ability to express ourselves in art—in pictorial language—preceded the ability to express ourselves in spoken language.*[139]

Lacking true language, and prior to pictorial images, our pre-Sapiens ancestors in Africa and Eurasia around 500,000 years ago probably communicated with one another by gesture and mimicry. This view is consistent with anthropologist records revealing that bands of hunter-gatherers communicated by means of gestures, facial expressions and mime.[140]

In contrast with tool-making, innovation in the human visual art tradition exploded in a very short and recent period. The earliest forms of pre-figurative decorative design reach back to about 140,000 years ago.[141]

Although rare evidence of symbolic behavior can be traced back to a number of African sites about 100,000 years ago, these artistic expressions appear more as flickers of creativity with little evidence of sustained expression. It is not until about 40,000 years ago that complex and highly innovative cultures appear and include behavior that would be recognized as typical of modern humans today.

Many researchers believe that this explosion of artistic material is at least partly attributable to a change in human cognition—an ability to think and communicate symbolically or to memorize better. Some also theorize that expanding tribal sizes and more complex social structures played key roles.

Dating back about 35,000 to 40,000 years ago, the Upper Paleolithic Löwenmensch mammoth ivory figurine or Lion-man of the German Hohlenstein-Stadel cave is the oldest-known uncontested example of figurative art. Carved gouges indicate that it was carved using a flint stone knife.

Stone-sculpted Venus figurines which appeared across Europe, from Portugal to Russia, between 28,000 and 21,000 years ago all had a remarkably similar style with exaggerated breasts. They are likely to reflect culturally-connected fertility symbols or mother goddess references.[142, 143]

Although decorative marine shell beads found in Israel date back about 90,000 years ago, and ostrich shell beads found in Morocco date back about 80,000 years, items of personal adornment (which were not sewn into clothing) only became prolific about 35,000 years ago. This suggests that human culture had come to attach growing importance to visual symbols of social appearance and status.

Musical instruments also began to appear around this time. Decorated mammoth bones shaped into castanets and flutes were found at various French Paleolithic Cro-Magnon sites ranging from 30,000 to 10,000 years old.[144]

Ever since the beginning of the cognitive revolution about 70,000 years ago, Sapiens have thus been living in a dual reality. One reality is directly observable and experienced, an objective reality of rivers, trees and lions. The other is an imagined reality, one of gods, laws and tribal customs.[145]

Visual arts that soon followed reveal an apparent merging of real-life story-telling with afterlife imaginings. Paleolithic paintings dating about 15,000-20,000 years ago in the Lascaux Cave in the Dordogne region of southwestern France depict what appears to be a man with the head of a bird and erect penis being killed by a bison. Another bird beneath the figure might possibly be interpreted to symbolize the soul released from the body at the moment of death.

Wall paintings Las Cueva de las Manos (Cave of the Hands) in the valley of the Pinturas River, in Argentina, date back roughly 10,000 years. Although there are three distinct styles which are believed to have been created at different time periods, the highlight is the hundreds of colorful handprints dated to around 5,000 BC.

It's believed these cave dwellers stenciled their own hands using bone-made pipes to create the silhouettes of their left hands, indicating that they probably held the spraying pipe in their right hands. Various mineral pigments were used to make different colors—iron oxides for red and purple, kaolin for white, natrojarosite for yellow and manganese oxide for black.

The cave also contains hunting scenes and representations of animals and human life dating back further to around 7300 BC. The hunter-gatherers who lived in the caves depicted pursuit of prey including the guanaco, a type of llama using the bola—cords with weights on either end which are thrown to trap the legs of the animal. A third category of paintings depicted animals and humans in a more stylized and minimalist fashion, done largely in red pigments.

Cognitive fluidity which emerged during the Upper Paleolithic period roughly 50,000 years ago enabled our ancestors to advance tool technologies, cooperative game hunting, tribal societies and ceremonies, language and art—an explosion of ever-accelerating creative progress which now grows at an exponential rate.

Language, including visual art, led to innovative story-telling lessons and mythologies which established and passed on post-generational societal mores, governance and status hierarchies, traditions and rituals and behavioral expectations and enforcements. These developments, in turn, strengthened cultural bonds, motivated cooperation and spurred more innovation through swarm intelligence.

Human brain plasticity provides a constantly updatable and expanding open-ended cognitive capability to synthesize and act upon observations, musings and theories to innovate new or modified insights. This mental agility enables expansive dimensions of innovative thinking which other animals lack.

The brain's plasticity feature supports what we recognize as emotional intelligence from which higher order evolution of social and adaptive problem-solving emerged. Our brains accomplish this versatility and flexibility through interconnected communications involving information processing centers which distributed throughout many locations.

Here, there may be a solid scientific basis to defend scatter-brained ideas after all. This creative network includes the emotional brain (limbic system), memory system and motor system that interact with the rational brain (neocortical deliberation system).

There is also real truth in the notion that creative people constantly reinvent themselves. As fresh experiences etch new neural pathways, plasticity enables our human brain to constantly reconfigure its synapse circuitry as an unceasing work in progress. Put quite simply, what fires together, wires together.

Innovations made possible through cognitive fluidity and plasticity were essential for the emergence of farming, which in turn enabled the emergence of specialists such as craftsmen and scribes necessary for the continued growth of intellectuality.

Eventually, academics, technologists and scientists appeared—fundamental to our exponential growth in creativity and knowledge today.[146]

The All-Important Domestication of Fire

The control of fire marked a major survival and cultural development in human evolution. Domestication of fire fundamentally influenced our Sapiens ancestors' geographic dispersal, food supply and diet, hunting weaponry and tools, art and utensils, safety from predators, societal cooperation and very possibly even their evolutionary physiology.

Evidence of widespread control of fire by anatomically-modern humans dates back as a gradual process to approximately 125,000 years ago. It likely began by transporting burning brush that had been ignited by lightning for heating and cooking purposes. Archaeological evidence suggests that Sapiens later figured out how to make fire with a drill friction device that produced ignition heat by hardwood rubbing against softwood.

Sophisticated control of fire provided a source of warmth which allowed tropical and subtropical Sapiens to survive severe climate changes, and to migrate into regions that even the cold-adapted Neanderthals had been unable to inhabit.

Additionally, fire allowed the expansion of human activity into the dark and colder hours of the evening, essentially increasing the length of daytime for social interactions. Whereas the modern human's waking day is typically about 16 hours, most mammals are only awake about half that many hours. The nighttime use of fire also served as a means to ward off dangerous nocturnal predators.

Hominids discovered that meat dried through the use of fire could be preserved for times when scarce game and harsh environmental conditions made hunting difficult. Fire also dramatically changed what and how food was consumed.

Before the advent of fire, the hominid diet was limited mostly to plants composed of simple sugars and carbohydrates, such as seeds, flowers and fleshy fruits. Cooking made starchy and fibrous foods edible and increased the diversity of foods that were available.

Over time, this development is theorized to have contributed a skeletal change. Prior to use of fire, hominid species had large premolars needed to chew harder foods such as seeds. In response to cooking, molar teeth of Homo erectus gradually shrunk. Accordingly, hominid jaw volumes decreased as well, adapting to a variety of smaller teeth.

Cooking also killed parasites, reduced the amount of energy required for chewing and digestion, and released more nutrients from plants and meat. This particularly benefited our Cro-Magnon ancestors when the frigid Ice Age climate compelled them to live more on meat, and less on plants.

The use of fire by early humans also served as an engineering tool to modify the effectiveness of weaponry. Archeological researchers excavating in an area known as the Spear Horizon in Germany unearthed eight 400,000-year-old fire-hardened wooden spears along with stone tools. One of the spears was found in the pelvis of a horse. A fire-hardened lance was found in the rib cage of a straight-tusked elephant at another dig site in Lehringen, Germany.[147]

More recent evidence dating to approximately 164,000 years ago found that humans living in South Africa in the Middle Stone Age used fire as a heat treatment for a fine-grained rock called silcrete. Once treated, the rocks were modified and tempered into crescent-shaped blades for arrowheads. This process was also likely used to create heat-modified tools for cutting meat of killed animals.

Fire has been used to create pottery dating back at least to evidence of 26,000-year-old high-temperature kilns and ceramic technology found at the Cro-Magnon site at Dolni Vestonice in the Czech Republic. The kilns were capable of firing clay figurines at temperatures over 400 degrees Celsius. About 2,000 fired lumps of clay were found scattered nearby.

Other 20,000-year-old pottery evidence was discovered in the Xianrendong Cave in China. However, it was during the Neolithic Age which began around 10,000 years ago that creation and use of pottery became far more widespread, a time generally associated with use in connection with early agriculture.

Beyond Hunters and Gatherers

The discovery and the use of fire, and the sharing of the benefits, may have created a sense of sharing as a group, for example, through the cooperative participation of gathering firewood. Ongoing cultural advancements which have enabled our very survival have also made us increasingly dependent upon vast networks of cooperation.

Social cooperation has been particularly important for Sapiens compared with other animals because in a cognitive development context our children are born prematurely. This being said, since humans are born underdeveloped, they can also be educated and socialized to a far greater extent than any other animal.

As Yuval Noah Harari notes:

> *[Whereas] most mammals emerge from the womb like glazed earthenware emerging from a kiln—any attempt at remolding will only scratch or break them. Humans emerge from the womb like molten glass from a furnace. They can be spun, stretched and shaped with a surprising degree of freedom.*

This being the case, lone mothers with inadequate time to forage for food for their offspring and themselves with needy babies in tow required constant help from other family members and neighbors.[148]

About 45,000 years ago humans were predominately nomadic hunter-gatherers living on a wide-ranging omnivorous diet of wild animals and plants. This menu versatility enabled us to utilize the food resources found in diverse environments.

Vigorously aggressive and incessantly waring, these small nomadic bands weren't behaviorally adapted to life in permanent communities—first clear evidence of a settled lifestyle was found in 18,000-year-old Ice Age mammoth bone houses in the eastern part of central Europe. Long-term settlement seems also to have occurred with Natufians in the Near East between 15,000 and 11,500 years ago.

By about thirteen thousand years ago, continental glaciation had begun to slowly retreat. The Pleistocene Ice Age epoch was replaced by a warm, dry period that necessitated adjustments to far more reliable food sources. As forests spread and big game inhabiting open plains became scarcer, fishing, bird and small animal hunting and gathering became increasingly vital.

Early settlement living would have depended heavily upon naturally abundant and reliable local food sources such as hazelnuts or salmon, along with methods of long-term storage. Although wild foods still remained important in the diet, it was only about 11,000 years ago that humans began to domesticate plants and animals.[149]

Events that promoted farming appear to have come about after settlements had come about and means of food storage had been invented. Subsequent storage surpluses led to trade between settlements, often utilizing river routes. Agricultural settlements dating back at least 9,600 years were established along the Euphrates River in northern Syria. Such trade moving by boats and along coastline routes promoted migration and population mixing.[150]

The end of the last Ice Age during the Upper Paleolithic period marked a brief Mesolithic period which witnessed the development of new tools, especially those for woodworking and fishing. This was soon followed by a new Neolithic culture of farming, a radical social transformation which began in the Middle East about 10,000 years ago and gradually spread through Europe.

Farming also began independently and at different times in Southwest Asia, Equatorial Africa, Southeast Asia, Central America and the lowland and highland South America. With the exception of the far north, agriculture had come to predominately replace hunting-gatherer lifestyles within two millennia.

A major transition to agriculture which between around 9,500-8,500 BC in the hill country of southeastern Turkey, western Iran and the Levant slowly spread in restricted geographical areas. Archeological records of early Middle East settlements, particularly those in the Levant, reveal that the Natufians populations (currently Israel) were then still essentially hunter-gatherers who subsisted on dozens of wild cereal species. Yet even at that early time, they had learned to build stone houses, to invent new tools such as stone scythes for harvesting wild wheat, to use stone pestles and mortars to grind it and to construct granaries to store food for future needs.[151]

Gradual change from hunter-gatherer to permanent settlements is evidenced by Natufian artifacts dating from about 10,300 BC to 8,500 BC. The oldest large settlement of about one hundred population was Jericho, circa 8,000 BC, whose merchants traded grain with Anatolia and Mesopotamia for obsidian, malachite and turquoise (named for importation from Turkey) as early as 8,300 BC.[152]

Natufian descendants gradually adopted important plant cultivation innovations. They began to lay aside some of the wild grains they gathered from harvests to sow and replenish plantings for the next season. They discovered that they could achieve better plant results by sowing the grains deep in the ground with hoes and ploughs than by haphazardly scattering them on the surface. They learned to weed fields to guard edible plants against parasites, and conceived ways to water and fertilize them.

Rivers were first used for irrigation in Jericho. By the sixth millennium BC, irrigation innovations centered on alluvial plains of Mesopotamia between the Tigris and Euphrates Rivers yielded the most productive agricultural area ever witnessed. This capacity to feed growing populations established locations for some of the world's earliest large cities.[153]

As cereal cultivation demanded effort which left less time to gather and hunt wild species, foragers transitioned to become farmers. Those tribes that resisted faced a serious dilemma.

Once agricultural settlements secured control, demographic growth in the area enabled farmers to overcome hostile challengers in small foraging bands by sheer weight of numbers. Their only remaining choices were to run away, abandoning their hunting grounds to field and pasture, or to take up the ploughshare themselves. In either case, their old lifestyle was forever over.

The first domesticated forms of Einkorn, a wild cereal, were developed about 10,500 years ago. Modern wheat arose about 7,000 years ago in Iran from a cross between a wild grass and a more recent emmer wheat domesticated in the Euphrates valley of modern Syria.[154]

Whereas 10 thousand years ago wheat was merely one wild grass among many, within a few short millennia it was growing all over the world. Wheat fields are now estimated to cover about 870,000 square miles of the globe's surface—nearly ten times the size of Britain.[155]

More than 90 percent of the calories that feed humans today came from a handful of plants that our ancestors domesticated between 9,500 BC and 3,500 BC. First were wheat, rice, barley potatoes and maize (called corn in the United States); followed by peas, lentils and potatoes around 8,500 BC; olive trees about 5,000 BC; and grapevines in 3,500 BC. Cashew nuts were domesticated somewhat later.

Concurrently, a new agricultural revolution progressed with the domestication of plants; our Sapiens ancestors also tailored to animals to serve desired purposes.

Dogs became human-domesticated partners about 15,000 years ago, likely supporting transitions from foraging to settled societies in a variety of ways. Canines could be trained to hunt, and their alertness to human and animal sounded barking sentry warnings. They served as welcome cold night bed warmers and provided a meat source in emergencies.

DNA records indicate that dogs were first bred from gray wolves which then rapidly spread rapidly throughout Eurasia. The earliest dog fossil dates back to 14,000 years ago in Germany.

Domesticated sheep, goats, cattle (from aurochs—an extinct breed of European cattle) and pigs (from wild boar) appeared between about 10,000 and 9,500 years ago. Horses were first domesticated on the Eurasian steppes around 6,000 years ago.[156]

The evolution of farming cultures introduced the domestication of animals in accompanying stages. Broadly summarized, this process likely began as nomadic Sapiens hunters first altered the constitutions of the herds through a process of selective hunting of older, sicker animals, sparing fertile females and youngsters to safeguard the long-term vitality of their food supply.

A second step may have been to actively defend herds from predators such as lions, wolves and even rival human bands. This would have included driving interlopers away, as well as corralling herds into a narrow gorge in order to better control and defend it.

The final big domestication stage involved a more careful selection among members of the herds in order to tailor them to preferred features. Those individuals that exhibited the most aggressive resistance to domestication control—along with those with skinniest meat content—were slaughtered first. Included by selective features were horses, cattle, donkeys, sheep, boars and chickens which provided food (meat, milk, eggs), raw materials (skins, wool) and muscle power (transportation, plowing, grinding and other physical tasks).

Animal domestication led to special new herding and agriculture societies needed to feed ever-growing populations. Since then, the few million sheep, cattle, goats, boars and chickens that existed in Afro-Asian niches ten thousand years ago has grown dramatically. Today's world contains about a billion sheep, a billion pigs, more than a billion cattle and more than 25 billion chickens...the most widespread fowl ever.

Early domestic farming experiments often led to big setbacks. Among these, diseases were inordinately potent consequences. Malaria is thought to have become common among humans 10,000 to 5,000 years ago in connection with West African introduction of slash-and-burn agricultural practices which left sunlit pools as breeding sites for Anopheles mosquitos. Other diseases transferred from livestock to people promoted infectious epidemics.

Archaeological bone and tooth evidence reveal that many of these early farming efforts also led to health problems not experienced by hunter-gatherers. Increased food production often couldn't keep pace with population growth. In particular, a lack of knowledge regarding fertilization and soil depletion degraded crop and livestock yields.[157]

Advancing settlement and farming lifestyles presented major challenges requiring new ways of thinking and social relationships.

Settlement living promoted an increasing need for social

cooperation commensurate with population densities. Founded on a principle of reciprocity, such cooperation not only served to enhance community prosperity and stability, but also promoted a social form of competitive natural selection that drove individuals and collaborating groups to achieve higher accomplishments.

Less fortunately, that increased social interdependence also gave rise to a new social problem whereby freeloaders and other cheats deceived others to further their own interests. Although human traits of cunning and deceit had long-existing history predating settlement cultures, close living and work-sharing circumstances required effective means to apply peer pressure—sometimes with very painful or mortal infraction penalties.

Whereas kinship was the primary tie among hunter-gatherer groups, increasing population densities required trading nomadic independence and equality for collaborative benefits of conformity within allegiance to governing domain rules and community hierarchy status.

Private property became an innovation that established class by different levels of ownership. Chiefs, commoners, rich and poor families, labor specialization and fixed social rules of behavior initiated life complexities that have continued to increase over time.[158]

The Development of Written Language Cultures

Conflicts arising from continuous, close contact and community defenses against marauding hunter-gatherers promoted the development of more and more elaborate writing communication methods essential for negotiation and collaboration.

Long-distance trade in Mesopotamia created a need to be able to communicate across the expanses between cities or regions. Written records used exclusively for such accounting purposes were a variety of more than 500 cuneiform or wedge-shaped pictogram symbols pressed into clay which served to aid in keeping track of such things as which parcels of grain had gone to which destination or how many sheep were needed for events like sacrifices in the temples. Many of these surviving records had to do with sales of beer, a popular beverage.

A type of phonetic writing which appeared after 2,600 BC was characterized by a complex combination of word-signs and phonogram—signs for vowels and syllables—that allowed the scribe to express ideas. Impressed on clay tablets, they were applied for a vast array of economic, religious, political, literary and scholarly documents.

The first alphabet which emerged in Mesopotamia around 2,500 BC was confined to consonants. The Greeks later created vowels the middle of the 8th century BC, completing an alphabet which continues today with only minor changes.

The first writer in history known by name is the Mesopotamian priestess Enheduanna, daughter of Sargon of Akkad, who wrote poems dedicated to the goddess Inanna.

Predating Homer and the historical part of the Bible, the first recording of major literature dates to 2,100 BC. Gilgamesh tells a saga of heroic King Ukruk and a flood which decimated Sumerian cities which has striking similarities to the Great Flood of Noah legend which was passed down to Assyrians and Hebrews. Evidence that such a flood occurred consists of a ten-foot deposit of silt at the location.[159]

Between 2,500 and 1,300 BC, Aryan warriors conquered most of India, merging their Semitic alphabet with Sanskrit of India to create new literary Brahmi script. Ancient Rig Veda Hindu poems and songs reveal a scarcely literate, egalitarian society prior to its takeover by a militaristic, patriarchic alphabetical one.[160]

Hindu Brahmins established strict laws forbidding contact with writing, and consequently, the Hindu civilization existed for two thousand years without a written law. Laws of Manu were later established following adoption of the Brahmi script.

Inventions of farming, language, the alphabet and writing catalyzed transformational new social arrangements, accountabilities, allegiances and alliances.

Writing's important contributions to these developments were many. It codified understandings between individuals, rules of behavior and penalties for noncompliance.

Writing provided a medium to document, disseminate and immortalize events and ideas for then-present and future generations (including us).

And, in various cases, both for better and worse, writing served as an instrument of power which was applied and exploited by social hierarchies and anointed leadership elites to exert dominant control. Insidious catch-22 consequences often resulted in the co-existence of cultural stability and advancement on one hand, and the concomitant promulgation of social exploitation, oppression and warfare on the other.

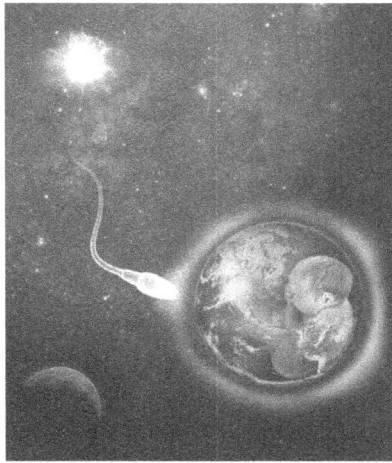

Part Six: The Rise of Empires and Societies

PRIOR TO THE agricultural revolution, those few human enclaves that existed were very small and surrounded by expanses of untamed nature. By the first century AD, only 1-2 million of the previous 5-8 million nomadic foragers still remained (mainly in Australia, America and Africa). Their numbers had become dwarfed by the world's 250 million farmers.[161]

The rapid expansion of permanent farming settlements along with trade and cooperative regional defense relationships created a need for more elaborate and absolute governance means to coordinate and enforce populace behaviors deemed mutually beneficial. Who actually deemed what was most beneficial, and beneficial to whom, varied greatly from one region to another, with great consequences to those governed or resisting sovereign authority.

A capability to feed a thousand or more people in the same town or region offers no assurances that individual members and groups within that population will easily agree how best to divide the land and water, how to settle disputes and conflicts and how to act together in times of drought or war. Common rules and penalties must also be established and enforced to ensure cooperative compliance.

The first millennium BC witnessed the appearance of three primary unifying orders of law and governance:

- The first, an economic order, perceived the known world as a single market, and all populations as customers.

- The second universal order was an imperial order, where conquerors saw the world their empire and all populations as their prospective subjects.

- The third universal order was religious, where prophets and believers viewed the entire world bound by particular articles of faith applicable to everyone everywhere such as Christianity and Islam.

The Universal Order of Money

The invention of money provided a universally trusted medium of cooperation and trade which enabled people almost everywhere to convert almost everything into almost everything else, while also providing a convenient means to store wealth and to transport it from place to place.

As Yuval Noah Harari contemplates:

> Imagine a wealthy farmer living in a moneyless land who emigrated to a distant province. His wealth consists mainly of his house and rice paddies. The farmer cannot take with him the house or paddies. He might exchange them for tons of rice, but it would be very burdensome and expensive to transport all that rice.[162]

Money's ability to convert, store and transport wealth easily and cheaply gave rise to complex commercial networks and dynamic trading markets. Cowry shells served this purpose for about 4,000 years all over Africa, South Asia, East Asia and Oceania. Yet like dollars today, their value existed only in common and imagined trust.

Harari observes that such trust in money involves and demands a very complex and long-term network of political, social and economic relations:

> *Why do I believe in the cowry shell or gold coin or dollar bill? Because my neighbors believe in them. And my neighbors believe in them because I believe in them. And we all believe in them because our king believes in them and demands them in taxes, and because our priest believes in them and demands them in tithes.*

Since trust is the raw material from which all types of money are minted, determinations of worth are often entirely independent of any inherent value attached to its structure or form. Still, gold, silver and other rare materials such as gemstones enjoyed early and far-ranging trade exchange appeal.

While there needed to be enough of the precious stuff available to accommodate trade demands, easy and reliable standards were also required to guarantee authenticity. Gemstones presented standardization and availability disadvantages for general use, and poorly served purposes of small value transactions. Precious metals were better but required precise weighing and verification of purity.

Set weights of precious metals eventually gave birth to coins which were first struck around 630 BC by King Alyattes of Lydia, in western Anatolia. Each coin had a standardized weight of gold or silver and was imprinted with an identification mark that testified to two things: the amount of precious metal it contained and the authority that guaranteed its contents.

Romans later developed the denarius coin which found popularity as a trade standard both within the empire and beyond. The coinage enabled the emperor to collect barley taxes in Syria without having to transport it in the form of grain to the central treasury in Rome, and then once again transfer the grain to Britain in order to pay its legions stationed there.

At its zenith, the Roman Empire collected taxes from up to 100 million subjects. That revenue financed a standing army of 250,000-500,000 soldiers, a road network still in use 1,500 years later and theaters and amphitheaters that host spectacles to this day.

During the first century AD, Roman denarius coins became an accepted medium of exchange in the markets of India thousands of miles away. In fact, the Indians had such confidence in the denarius that even coins they struck there retained the Roman emperor's image.

Denarius became broadly adopted as a generic term for coins. Muslim califs Arabicized this to dinars, a term still applied to currencies used in Jordan, Iraq, Serbia, Macedonia, Tunisia and several other countries.[163]

Currency trading emerged as merchants traveling between India and the Mediterranean noticed differences in the value attached to gold. In order to make a profit, they would buy gold cheaply in India and sell it dearly in the Mediterranean. This caused the demand and value of gold in India to skyrocket, while at the same time, the Mediterranean would experience an influx of gold, with its value consequently dropping. Within a short time, the value of gold in India and the Mediterranean would be quite similar.

China developed a slightly different monetary system during the seventh century BC which was based upon bronze coins and unmarked silver and gold ingots. The entire world later relied heavily upon gold and silver along with a few trusted currencies such as the British pound and the American dollar.

The Universal Imperial Order

Whereas emerging cities and regional domains such as kingdoms began to cooperate with certain others outside, who they called brothers or friends, there was always an inevitable natural tendency to discriminate between the us versus them—those less worthy barbarians who resided in the next valley or beyond the far mountain range.

Deadly conflicts sometimes arose when those outsiders possessed fertile land and material wealth coveted by insiders or their leaders. A popular solution was to establish more powerful empires needed to conquer them and either make them part of "us"—or alternatively, to either enslave or exterminate "them" This occurred, for example, when Romans who invaded Scotland in 83 AD encountered fierce resistance from local Caledonian tribes and reacted by laying waste to the country.

And sometimes populations who actually were conquered eventually ceased to view the empire as an alien system of occupation whereby the ruled and rulers alike viewed the other them as a new us. After centuries of imperial rule, all Roman subjects were granted citizenship. Some non-Romans even rose to top ranks in the officer corps of the Roman legions.

Other non-Romans were appointed to the Senate. In 48 AD, the emperor Claudius noted in a speech that several Gallic Senate appointees who through "customs, culture, and the ties of marriage have blended with ourselves."

Powerful empires need not necessarily emerge from military conquest. The Athenian Empire began as a voluntary league, and the Hapsburg Empire came about through a string of mutually beneficial marriage alliances.

Nor must an empire be ruled by autocratic leaders. The

empires of Novgorod, Rome, Carthage and Athens were governed as democracies, as was the enormous British Empire of modern times.

Size doesn't always determine empire status either. The Athenian Empire at its zenith was much smaller in size and population than today's Greece, and the Aztec Empire was smaller than today's Mexico. Yet while both of these were true empires which gradually subdued dozens, and even hundreds, of different polities, modern Greece and modern Mexico, which lack such diversity, do not qualify. Athens once presided over more than a hundred formerly independent city-states, whereas the Aztec Empire ruled over 371 different tribes and peoples.[164]

Fundamentally, an empire is a political order with two important defining characteristics. First, it must rule over a significant number of distinct peoples, each possessing a different cultural identity and a separate territory.

Second, empires are characterized by flexible borders and an ability to swallow and digest more and more additional nations and territories without altering their basic structure or identity. Together, cultural diversity and territorial flexibility give empires not only their unique character, but also their central role in history.

Some empires were clearly much mightier than others. Some last for thousands of years, while others last only months. Some stretched across multiple continents, while others only existed on one single continent. And some empires were more savage than others, depending upon numerous wars and brutal punishments to maintain their power structures.

The inventive and enterprising Sumerians of ancient Mesopotamia who built canals, dams and reservoirs to support Mesopotamian agriculture and human sustainment before 4,000 BC also created violent empires headed by cunning rulers wielding absolute power.

During the fifth and fourth millennia BC, cities with tens of thousands of inhabitants sprouted in the Fertile Crescent, a quarter-moon-shaped region from the Persian Gulf through modern-day southern Iraq, Syria, Lebanon, Jordan, Israel and northern Egypt. Each of these cities held sway over many nearby villages. The rapid growth of city-states incited a mad scramble for wealth. Empires came into being in response to boundary disputes over irrigated lands and control of valuable resources including metals, stone and timber.

Among multiple conflicts, the first war in 2,525 BC between Umma and Lagash, two cities nearly 18 miles apart, was over the fertile region of Guendena. Remnants of memorials commemorate the victory over the king of Umma at the hands of Eannatum of Lagash. Such defeats typically ended badly for captured opponents who were carted away as slaves.

Around 2250 BC, Sargon the Great forged the first true empire, the Akkadian, with over a million subjects and a standing army of 5,400 soldiers. Sargon began his career as a gardener, and then as King of Kish, a small city state in Mesopotamia. Within a few decades he managed to conquer not only all other Mesopotamian city-states, but also captured large territories outside the Mesopotamian heartland and covered much of modern-day Iraq.

After the Akkadian empire fell, it was succeeded by two great mega-empires with millions of soldiers—Babylonia to the south, and Assyria to the north. Assyrians were known to threaten their enemies and captives with savage brutality. Two thousand years later when the Assyrians were defeated, its nobles were massacred in revenge for their tyranny.

In 3100 BC, the entire lower Nile Valley was already united into an Egyptian kingdom, where pharaohs came to rule an empire covering thousands of square miles and ruling hundreds of thousands of people. Unlike other major empires, Egypt remained stable without warfare for nearly three thousand years. This stasis ended with the Roman conquest of Egypt in 30 BC.

Surrounded on three sides by deserts, the Isthmus of Suez and the Mediterranean on the remaining side, Egypt's borders had previously required little defense. Consequently, when the first pharaoh, Menes, united Egypt around 3,000 BC, the alien "barbarians" beyond those natural borders presented little interest other than to the extent that they had land or natural resources that the pharaoh wanted.

Avoiding great conflicts with neighboring empires and supported with fertile agriculture, the pharaohs turned priority attention, power and wealth in preparing for their own afterlives. Historian Charles Van Doren theorizes that this may explain why other than great feats of engineering exemplified by the Great Pyramids, Egypt's technological progress advanced less rapidly than in many other empires. In other words, invention wasn't essential for survival or better living.[165]

In 221 BC, the Qin dynasty united China, where taxes levied on 40 million Qin subjects paid for a standing army of hundreds of thousands of soldiers, and a complex bureaucracy that employed more than 100,000 officials. Shortly afterwards, Rome united the Mediterranean basin.

The Mongol Empire which existed during the 13th and 14th centuries emerged from the unification of several nomadic tribes under the original leadership of Genghis Khan. The largest contiguous land empire in history, its territory eventually stretched from Eastern Europe and parts of Central Europe to the Sea of Japan, extending northwards to Siberia, eastwards and southwards into the Indian subcontinent, Indochina and the Iranian Plateau, and westward as far as the Levant and the Carpathian Mountains. Connecting the East and the West, it allowed the dissemination and exchange of trade, technologies and ideologies across Eurasia.

Any resistance to Mongol rule was met with massive collective punishment. Before conquering a city, they would send a messenger to demand surrender. No harm would come to those who did chose to surrender but unbelievable brutality would meet those who did not. In 1258 AD when Baghdad refused to surrender to Genghis Khan's grandson Hulagu Khan, its population was massacred, the city was sacked and its libraries of rich information were destroyed. As a special warning to others who might consider defying Mongol orders, the city ruler was wrapped up in a rug and beaten to death.

The Mongol populace was governed by a code of law called Yassa, meaning order or decree, with strictly-enforced canons carrying severe penalties. For example, it decreed a death penalty if one mounted soldier following another did not pick up something dropped from the mount ahead of it, and also meted out harsh responses to rape and murder.

The Yassa guaranteed its subjects some important freedoms as well. Chiefs and generals were selected based upon merit, those of rank shared much of the same hardships of the common man, and virtually every religion including Buddhism, Christianity, Manichaeism and Islam were assured freedom to be practiced, and all religious leaders were exempt from taxation and public service.

By 1450 AD, close to 90 percent of humans lived in a single mega-world: the world of Afro-Asia. Most of Europe, and most of Africa (including substantial chunks of sub-Saharan Africa) were then already connected by significant cultural, political and economic ties.

Most of the remaining tenth of the world's population was divided between four worlds of considerable size and complexity:

- The Mesoamerican World, which encompassed most of Central America and parts of North America.

- The Andean World, which encompassed most of South America.

- The Australian World, which encompassed the continent of Australia.

- The Oceanic World, which encompassed most of the islands of the south-western Pacific Ocean, from Hawaii to New Zealand.

Over the next 300 years, the Afro-Asian giant swallowed up almost all the other worlds. It consumed the Mesoamerican World in 1521, when the Spanish conquered the Aztec Empire. At about that same time it began to overtake the Oceanic World during Ferdinand Magellan's circumnavigation of the globe, then soon afterwards completed that conquest. Spanish conquistadors advanced to crush the Inca Empire in 1532, collapsing the Andean World.

After the first European landed on the Australian continent in 1606, British colonialism began in earnest in 1788. Fifteen years later the Britons established their first settlement in Tasmania, thus bringing that last autonomous human world into the Afro-Asian sphere of influence.

The Universal Order of Religion

Alongside money and empires, religion has constituted the third great unifying (and dividing) influence of humankind. Whereas it has served as the glue that binds social stability to deal with common threats, it has also functioned as a cohesive medium to secure rule by tyrants.

Ruling submissive subjects through violence and cunning, some sovereigns asserted their own personal supreme divinity status. Others enlisted alliances with prophets and priests to impose strictures of despotic religious authority.[166]

Broadly defined, the term religion pertains to a system of human norms and values that is founded upon a belief in some form of superhuman order. Further, a religion typically asserts which particular norms and values are to be considered most important and binding as ordained by an absolute and supreme authority.

What might be described as a religious revolution might trace back to the transition from hunters-gatherers to farmers at the early beginnings of the agricultural revolution.

Any sense of supremely-ordained human status might not have readily occurred to those early ancestors who spent their lives picking wild plants and pursuing wild animals they regarded to be natural equals. To them, the fact that they hunted sheep did not make sheep inferior to them, just as the fact that tigers hunted them did not make them inferior to tigers.

This likely began to change after people domesticated some of those sheep and became responsible for keeping their now-human-dependent flocks safe from other human and animal predators and healthy from disease epidemics arising from closer confinement.

The help of deities such as a fertility goddess, sky god and god of medicine might first have been called upon to mediate between humans and their newly co-dependent plants and animals. Religious liturgy for thousands of years after the agricultural revolution focused heavily on sacrifices of lambs, wine and cakes to divine powers in exchange for fecund flocks and abundant harvests.

Chronic, often bloody, ideological conflicts later arose regarding whether one particular god or a more specialized power-sharing group of gods wielded divine supremacy. Those subscribing to a single god supremacy (or theocracy) are termed monotheists. Other faiths that understand the world to be controlled by a group of powerful gods are known as polytheistic religions (from the Greek "poly" = many, "theos = god).

The first known monotheist religion appeared in Egypt around 1,350 BC when Pharaoh Akhenaten declared that the god Aten (but one of several minor deities of the Egyptian pantheon) wielded supreme power over the Universe. Worship of Aten over all other gods became institutionalized as the state religion. Akhenaten's religious decree ended with his death when the worship of Aten was abandoned in favor of the old pantheon.

Monotheists have tended to be far more fanatical and missionary than polytheists. This behavior arises from a prevalent belief that since they are in possession of the entire message of the one and only true God, they are therefore compelled to discredit all other religions. This is evidenced by countless violent wars over the last two millennia as monotheists have repeatedly tried to exterminate all religious competition.[167]

As noted by Yuval Noah Harari, the monotheists are clearly winning. At the beginning of the first century AD, there were hardly any monotheists in the world.

Throughout its history, Judaism has not been a missionary religion. Yet through missionary zeal, a small Jewish sect took over polity of the mighty Roman Empire and established a spiritual Catholic headquarters in the capital city. Within the next 500 years, missionaries were busy spreading Christianity to other parts of Europe, Asia and Africa.

Rapid Christian expansion served as a precedent for another monotheist religion that appeared in the Arabian Peninsula in the seventh century—Islam. Like Christianity, Islam began as a small sect led by a charismatic prophet, spread throughout vast lands, and came and continues to play a central role in world history.

Born in Mecca around 570 AD, Muhammad, the Prophet of Islam, was orphaned at age six, raised by his uncle Abu Talib. Working as a trade merchant between the Indian Ocean and the Mediterranean Sea, he became greatly respected for trustworthiness.

At the age of about 40, Muhammad began to experience visions and repeatedly sought solitude in a cave on Mount Hira on the outskirts of Mecca. It was there that an angel, Jebreel (Gabriel), appeared to him and revealed messages from the one true God, Allah, which are recorded in the Quran.

In the year 610 Muhammad began preaching publicly, proclaiming that "God is One." Although he slowly won over a small group of ardent followers, including his wife, repeated efforts to attract Jews to his cause were unsuccessful.

Muhammad's monotheistic teachings which impugned the traditional polytheistic worship of gods and goddesses provoked powerful merchants and other Meccan city leaders to persecute him. He and his small band of followers were ultimately forced out of the city in 622 AD. The group established new residence in Medina (then known as Yathrib) where Muhammad united tribes under a Constitution of Medina and gathered together an army of about 10,000 converts.

In 628, Muhammad negotiated a truce with the Meccans, and in the following year, returned as a pilgrim to the city's holy sites. However, the murder of one of his followers provoked him to march upon the city, which soon surrendered. There was little bloodshed, with Muhammad only demanding of the Meccans that pagan idols be destroyed.

Muslims continue to revere Muhammad as the final prophet of God and the words of his messages recorded in the Quran are regarded as ultimate truths.

The world of Islam rapidly expanded as Egyptians, Syrians and Mesopotamians came to increasingly be dominated by non-Arab Muslims, in particular by Iranians, Turks and Berbers. Already by the time of Muhammad's death in 632, most of the Arabian Peninsula had been converted to the religion.

By the end of the first millennium AD, most people in Europe, West Asia and North America were monotheists, and empires from the Atlantic Ocean to the Himalayas claimed to be ordained by the single great God.

By the early sixteenth century, monotheism dominated most of Afro-Asia, with the exception of East Asia and the southern parts of Africa. It then began to gain dominance moving towards South Africa, America and Oceania. Most people outside East Asia now adhere to one monotheist religion or another, and the entire global political order is built on monotheist foundations.[168]

Nevertheless, many polytheists also essentially subscribe to a belief in a supreme God power, albeit one that presides over all lesser gods. In classical Greek polytheism, for example, Zeus, Hera, Apollo and their colleagues were subject to an omnipotent and all-encompassing power—Fate (Moria, Ananke).

The Hindus built no temples to their supreme God, Atman, for the same reason that the Greeks built none to Fate. Rather than seeking assistance from Atman in addressing such local worldly matters as winning wars and healing illnesses, these fall more into the domain of more specialized deities specialized such as Ganesha, Lakshmi and Saraswati.

Here, Fate and Atman are is too busy dealing with big Universe issues to be concerned with mundane desires, cares and worries of humans. From such an all-encompassing spiritual vantage point, it makes no difference whether a particular kingdom wins or loses, whether a particular city prospers or withers or whether a particular person recuperates or dies. So, from this perspective, it's pointless to ask that ultimate power over all for preferential victory in war, for health or for rain.

Accordingly, the Greeks made no religious sacrifices to Fate, and the Hindus built no temples for Atman. And while the Aztec Empire obliged its subjects to build temples for the lead God Huitzilopochtli, they were frequently constructed alongside those of local gods, rather than in their stead. Many of the imperial elite honored those polytheistic gods and rituals.

Unlike monotheists, polytheistic empires generally didn't attempt to convert subjects of new territories they conquered or coveted. The Egyptians, the Aztecs and the Romans didn't send missionaries to foreign lands to spread the worship of Osiris, Huitzilopochtli or Jupiter—nor dispatch armies to enforce any such intent.

Early Romans readily adopting foreign deities such as the Asian goddess Cybele and the Egyptian goddess Isis to their pantheon. The monotheistic evangelization of Christianity presented a long-resisted special problem in this regard—one seen as disrespect for the empire's protector gods, disloyalty to the emperor and ultimately, a politically subversive danger.

Persecution by polytheistic Romans led to thousands of tragic Christian executions over the over the course of three decades. Still, these casualties were dwarfed over the course of the next 1,500 years by millions of Christians slaughtered by other Christians in theological disputes over slightly different interpretations of that same religion. Hundreds of thousands of Catholics and Protestants killed each other during just the sixteenth and seventeenth centuries alone.[169]

Paradoxically, some ardent monotheistic religions, including Christianity, also sanctify de facto polytheistic beliefs and liturgies. Just as Jupiter defended Rome, and also as Huitzilopochtli protected the Aztec Empire, every Christian kingdom has enlisted divine help and protection from a particular patron saint assigned to their special causes.

England was protected by St. George, Scotland by St. Andrew, Hungary by St. Stephen and France had St. Martin. Cities and towns had their own saints also: Milan had St. Ambrose, while St. Mark watched over Venice.

Even various professions have had their own saints. St. Florian protected chimney cleaners, whereas St. Mathew lent a hand to tax collectors in distress. If you suffered from headaches you had to pray to St. Agathius, but if from toothaches, then St. Apollonia was a much better choice.[170]

An altogether different kind of religion altogether began to spread through Afro-Asia in the first millennium BC. These creeds, including Jainism and Buddhism in India, Daoism and Confucianism in China and Stoicism, Cynicism and Epicureanism in the Mediterranean basin put supreme faith in natural laws. While some of these religions continued to espouse the existence of gods, these deities were subject to obeying the same laws that govern humans, animals and plants.

The central figure in Buddhism, the most influential of these ancient, natural-law religions, is not a god, but a human named Siddhartha Gautama. Born around 500 BC as prince heir to a small Himalayan kingdom, Gautama was affected by suffering evident all around him. He observed that even those who were rich and famous are rarely satisfied with their lives.

Prince Gautama traveled as a homeless vagabond throughout northern India over six years, realizing in the process that suffering is not caused by ill fortune, by social injustice or by divine whims of gods. Rather, suffering and discontentment arise from behavior patterns of one's own mind.

Gautama established a set of ethical rules applying meditation techniques that focus attention to high ethical principles and away from cravings for power, sensual pleasure or wealth. He taught that perfection of these spiritual attainments leads to nirvana (literally meaning to extinguish the fire).

According to Buddhist tradition, Gautama himself attained nirvana and was fully liberated from suffering. Henceforth, he became known as Buddha, which means The Enlightened One.

The terms Hindu or Hinduism denote an unbounded set of theological beliefs with no compulsory dogmas that have persisted for nearly 4,000 years. Predicated on the idea that eternal wisdom of the ages can't be confined to a single sacred book, it embraces doctrines and practices ranging from pantheism to agnosticism, from faith in reincarnation to belief in the caste system.

Whereas other religions may look to find God in the heavens, the Hindu looks within themselves. There is no Hindu pope; no Hindu Vatican, no Hindu catechism; no prescribed divinities to adore or pray to; no religious Hindu rituals to honor or customs to practice; and no visible sign of Hindu identity to wear or display.

Given that Hindus make no common claim regarding what

God looks like, individuals are free to imagine Him or Her as a woman with eight arms riding a tiger; as a pot-bellied man with an elephant's head; or as a muscular figure with a monkey's head and tail.

Hindu texts operate upon a foundation of skepticism regarding heavenly certitude dating back to a compilation of 3,500-year-old Rig Veda poems. The hymn *Nasadiya Sukta* addressing mysteries of creation concludes: "In the highest heaven, only He knows—or perhaps he does not know."

Whereas most faiths believe that the soul has a body, most Hindus believe that the soul has a temporary body which the Eternal Atman resides in, then discards. Some others dismiss any need to accept belief in the existence of any God, whatsoever.

Hinduism is perhaps the single major religion not claiming to be the only true religion. As declared by the noted 19th-century Swami Vivekananda, Hinduism teaches not only tolerance of other faiths, but acceptance of them as well.

Believers are encouraged to find their own answers to the true meaning of life through emphasis on the mind, valuing reflection, intellectual inquiry and self-study.[171]

Establishment of Moral Principles and Laws

English novelist Aldous Huxley argued that there is a perennial philosophy or core of moral principles that exist in every time and place throughout history. Although variously interpreted through contemporary religious mandates and societal laws, some basic moral scaffolding appears to have been constructed upon common ancient foundations.

The Code of Hammurabi is one of the oldest recorded written laws containing a collection of 282 rules established by Babylonian king Hammurabi who reigned from 1792 to 1750 BC. Two earlier but less famous codes of conduct from the Middle East bear strong similarities: one created by the Sumerian ruler Ur-Nammu in the 21st century BC; the other a Sumerian Code of Lipit-Ishtar two centuries before Hammurabi.

Hammurabi's Code, which was carved onto a massive black stone pillar, codified principles of justice and punishment which are both prescient and bizarre by today's standards. In the former category is a doctrine of innocent until proven guilty, although with the extreme misfortune to any accuser who failed to make their case. Translated, it says, "If anyone brings an accusation of any crime before the elders, and does not prove what he has charged, he shall, if it be a capital offense charged, be put to death."

Another famous example is the ancient precept of retaliatory justice commonly associated "an eye for an eye," one often leading to grisly consequences. If a man broke the bone of one of his equals, his own bone would be broken in return. And sometimes there was far less equality in extracting reciprocal retribution for an offense. If a pair of scheming lovers conspired to murder their spouses, both were impaled, and if a son hit his father, the code demanded that the boy's hands be hewn off.

The severity of a penalty depended upon the social class level of the lawbreaker and victim. For example, If a man knocks out the teeth of his equal, his teeth shall be knocked out, whereas committing the same crime against a member of a lower class was punished only with a fine. If a man killed a pregnant maid-servant, he was punished with a monetary fine, but if he killed a free-born pregnant woman, his own daughter would be killed. Men were allowed to have extramarital relationships with maid-servants and slaves, but philandering women were to be bound and tossed into the Euphrates along with their lovers.

Hammurabi's Code also mandated a pecking order of appropriate compensation standards for payment of services based on social class and occupation. A doctor's fee for curing a severe wound was set at 10 silver shekels for a gentleman, five shekels for a freedman and two shekels for a slave. Penalties for malpractice followed the same principle: a doctor who killed a rich patient would have his hands cut off, while only financial restitution was required for maltreatment of a slave.

The code also specified minimum wages. Field workers and herdsmen were guaranteed a wage of eight gur of corn per year, whereas ox drivers and sailors received six gur.

Hammurabi's code remained influential in the region for centuries after his empire went into decline after his death in 1750 BC and crumbled completely when a Hittite army sacked Babylon in 1595. In the words of his monument, the purpose of his code was "to prevent the strong from oppressing the weak and to see that justice is done to widows and orphans."

According to one ancient Hindu creation myth, the gods fashioned the world out of the body of a primeval being (a cosmic man whose sacrifice by the gods created all life)—the Purusa. The sun was created from the Purusa's eye, the moon from the Purusa's brain, the Brahmins (priests) from its mouth, the Kshatriyas (warriors) from its arms, the Vaishyas (peasants and merchants) from its thighs and the Shudras (servants) from its legs.[172]

In *Songs of Chu*, an anthology of Chinese poetry during the Han dynasty (340 BC- 278BC), when the goddess Nüwa created humans from earth, she kneaded aristocrats from fine yellow soil, whereas commoners were formed from brown mud.[173]

Numerous philosophers and theologians have identified key articles of belief they believe to serve as a guiding universal set of moral values throughout history. Richard Kinner and Jerry Kernes in the Division of Psychology in Education at Arizona State University, Tempe, along with Phoenix counselor Therese Dautheribes, compiled a list of priorities which drew upon well-known texts and documents of major world religions.

The research team's document review included examinations of Judaism (the *Tanaka*), Christianity (the *New Testament*), Islam (the *Quran*), Hinduism (the *Upanishads* and the *Bhagavad Gita*), Confucianism (the *Analects of Confucius*), Taoism (the *Tao Te Ching of Lao Tzu*) and Buddhism (the *Dhammapada*).

The study also consulted with and reviewed materials of several secular organizations, including the American Atheists Inc. (*Atheist Aims and Purpose, Atheism Teaches That*, and *Introduction to American Atheists*), the American Humanist Association (*Humanist Manifesto I, 1933* and *Humanist Manifesto II, 1973*) and the United Nations (*Declaration of Human Rights*).

Altogether, here's what they came up with:

Commitment to something greater than self:

- To recognize the existence of and be committed to a Supreme Being, higher principle, transcendent purpose or meaning to one's existence.

- To seek truth (or truths).

- To seek justice.

Self-respect, but with humility, self-discipline and acceptance of personal responsibility:

- To respect and care for oneself.

- To not exalt oneself or overindulge—to show humility and avoid gluttony, greed, or other forms of selfishness and self-centeredness.

- To act in accordance with one's conscience and to accept responsibility for one's behavior.

Respect and caring for others (i.e., the Golden Rule):

- To recognize the connectedness between all people.

- To serve humankind and to be helpful to individuals.

- To be caring, respectful, compassionate, tolerant and forgiving of others.

- To not hurt others (e.g., do not murder, abuse, steal from, cheat or lie to others).

Safeguard the natural world:

- To respect other living things.
- To protect the environment.[174]

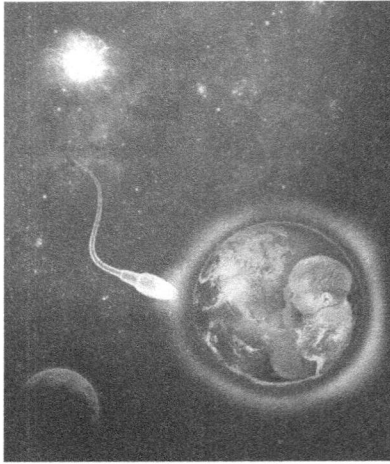

Part Seven: Fitful Starts and Struggles for Understanding

TRANSITIONS FROM AND between dogmatically metaphysical and empirically rational references of thought and behavior have been stutteringly erratic and turbulently brutal. Remarkably rapid intellectual advancements which began in Western Europe during the early centuries preceding the birth of Christ were fiercely resisted by religious hierarchies heading the early Church named in his honor.

Cruel penalties were inflicted upon those whose ideas were perceived to conflict with sanctioned articles of faith and dictates of all-powerful pontiffs. Bloody "holy" Crusade conflicts between the Church and non-compliant ideologies—as well as wars between various Christian sects—made free-thinking a perilous offense.

A Church-orchestrated and effectuated Inquisition exacted unthinkable terrors against those suspected of heresy. Included were young female children accused of witchcraft who suffered particularly gruesome fates.

Church corruption and populace control gradually gave way to a Renaissance emergence from a dark quagmire of anti-intellectualism that had smothered cultural and scientific advancement over nearly 200 years up until the beginning of the European High Middle Ages around 1000 AD.

Although empiricism continued to struggle against mysticism, there was and is no way to stop its determined strides of expansion and achievement. Followed and fueled by the Scientific Revolution and period of Enlightenment, civilization had chartered an open pathway to unlimited discovery, creativity and individual potential.

The endless expansion of human innovation, of course, long-predates medieval minds and European heritage. Striking creative advancements are evident in tool-making during the Upper Paleolithic period which began about 40,000 years ago, the invention of agriculture in the Neolithic about 10,000 years ago and early Harappan urbanization which peaked in India about 2,000 years ago.

Chinese and Egyptian Roles in Paving European Progress

Much progress which laid important paving stones for European scientific and cultural advancement can be credited to Chinese origins. Particularly consequential among these were contributions to horticulture, mechanical printing, means for precise ship navigation and gunpowder.

Chinese farming in southwest Asia's Fertile Crescent led about 10,000 years ago to the invention of grafting. They learned that trees grown from special cuttings produce reliably higher quality fruits such as apples, pears and cherries than those grown from seeds. This discovery contributed to an ability to feed their world's largest population.[175]

Also notable, it was the Chinese who invented movable

type, not Guttenberg who later developed the process in Europe where it had a greater revolutionary impact. Chinese applications of this printing method were made far more difficult due to an ideographic picture writing system involving myriad visual concepts rather than a comparatively very short and simple alphabetic iconography.

In combination with advanced printing technology, the Chinese also invented paper to go with it.

The Chinese discovered the circulation path of blood before English physician William Harvey, who is broadly credited for this, later did so in the 16th century. They also described the scientific first law of motion before English physicist and mathematician Sir Isaac Newton formulated quantitative proofs in the 17th century.[176]

Chinese invention of magnetic compasses and mechanical clocks greatly influenced world exploration. Seafarers prior to these innovations navigated by dead reckoning and coastal guidance. For latitude, they relied upon sighting of the Pole Star elevation above the horizon, yet because it vanished at the horizon, then used the Sun's position in the southern hemisphere as a proxy. This presented a problem because the Sun's observed elevation shifted throughout the year, requiring calculation by charting tables.

Meanwhile, a more accurate location also required timing calculations of celestial movement astronomy in the longitudinal plane. Although early navigators correctly understood the time function, precise longitudinal measurements required non-existent accurate clocks.

Fifteenth-century Chinese explorers put these seafaring navigation advancements to ambitious use which arguably eclipsed later European voyages of discovery. Between 1405 and 1433 AD, Admiral Zheng He of the Chinese Ming dynasty led seven huge armadas from China to the far reaches of the Indian Ocean. The largest of these comprised almost 300 ships and carried close to 30,000 people.

Zheng He's sailors visited Indonesia, Sri Lanka, India, the Persian Gulf, the Red Sea and East Africa. His ships anchored in Jeddah, the main harbor of the Hejaz, and in Malindi, on the Kenyan coast. Christopher Columbus' fleet of 1492 consisting of only three small ships manned by 120 sailors was very tiny by comparison.

During the 1430s, a new ruling faction of Beijing overlords abruptly terminated Zeng He's operations. His great shipping fleet was dismantled, crucial technological and geographical knowledge was lost and no explorer of such stature and means ever set out again from a Chinese port.

The Chinese invention of gunpowder later had consequential world-wide military influences upon large and small feudal systems and empires alike. Together with horticultural advancements to support ever-growing populations; machine printing and paper, which extended public information access; the clock, which enabled precise navigation and changed the broad ordering of human activities, it impacted civilization in immeasurable ways.

Yet China's influence would have been much greater were it not for the country's voluntary withdrawal from the outside world and their leaders' suppression of free inquiry. Reliance upon mystical texts aggravated by ideographic written language and bureaucratic suppression of scientific inquiry between the 15th and 20th centuries contributed to China's demise as an invention capitol.[177]

China's example once again demonstrates the pervasively inhibitive influences of powerful hierarchical bureaucracies. Overseen by strong centralized authority, a huge civil service had been established to control the country's vast irrigation system.

Promoting severely stratified society, the system discouraged efforts to grow to high achievement levels. To this day, the East evidences far higher conformity to tradition and authority than does the West.[178]

From Christ's time to the mid-14th century, the Chinese were prolific inventors, yet ultimately lagged the West in developing science. Here, more abstract ancient Greek thinking corresponded more closely with creative incentives for theoretical scientists striving to advance knowledge beyond desolate frontiers of bureaucratic and dogmatic ignorance.

Critical thinking, which relegated ancient Greek gods to the background, also contrasted with conditions in Egypt. There, the most learned men became priests who suppressed seminal thinking which their orthodoxy considered to be heretical. Egyptian adherence to sacred beliefs, subservience to authority and commitment to major works of cooperation exemplified by pyramid construction prevailed, whereas ancient Greek efforts became more individualistic.

Nevertheless, the Egyptian industry benefited Greek intellectuals as well. By the middle of the 8th century BC, the ancient Greeks imported papyrus from Egypt. This transportable, writing substrate enabled the Greeks to more readily distribute commercial records and treatises on technical matters.

Surges of Creative Greek Genius

From about 2000 BC until the 2nd century AD, ancient Greece became a primary center of intellectual achievement. This began as Greek-speaking Bronze Age nomads from distinct but culturally-connected regions invaded lands of the Aegean in intermittent waves.

The new settlers gradually adopted a new sense of social consciousness which questioned many previously fixed ideas. Although continuing to observe traditional religious customs and fertility rites, acceptance of human sacrifice by the ceremonial killing of prisoners declined.

Homer had completed his epic *Iliad* and the *Odyssey* poems by 800 BC. They are now regarded as central works of ancient Greek literature. The *Iliad* theme is set during the Trojan War, the ten-year-long siege of the city of Troy by a coalition of Greek kingdoms.

Whereas Homer probed the past for knowledge, later Greek philosophers looked to the future.

Nevertheless, primitive thought survived in the form of 4th and 5th century BC mythologies such as one attached to the legendary prophet, poet and musician Orpheus, who descended into the Underworld and returned—the source of Greek Orphic literary tragedy.[179]

By the mid-450s BC, Athens had grown to become a thriving commercial, maritime city where contact with wide-ranging cultures and world perspectives advanced conceptual thinking. Pericles, a great Athenian statesman, is credited as a powerful force in establishing the city as an important political and cultural empire. The Parthenon's construction in 447 BC is emblematic of Athens' renowned democratic stature.

Greek creative explosion represents one of humanities' giant leaps in human cognition. Gradually replacing anthropomorphic deities with natural phenomena, ancient Greek philosophers gradually intertwined naturalism with mythological thought. The imagined powers of gods waned as religious skepticism began to pervade Greek intelligentsia.

Rationalism rapidly came to the Greek mind by emphases on debate, politics and scientific schools in Ionia. Although the Greek majority continued to respect Olympian gods, Ionian scientists devoted their thoughts more an emphasis upon material substance which possessed neither divine order nor purpose.[180]

The epicenter for such discourse and debate was near the Ionian-dialect-speaking Greek city of Miletus, which prior to 500 BC, had become a major trade outlet for products from the interior of Anatolia in western Turkey and Sybaris, in southern Italy.

Larry Bell

In addition to its commerce and colonization, Miletus was distinguished for its literary and scientific-philosophical figures. Revolutionary thinking seemingly occurred in Miletus with remarkable suddenness as Ionian philosophers replaced mythology with critical rationalism based upon speculations regarding natural phenomena as understood at that time.

Around 600 BC, Thales of Miletus made the seminal observation that the world possessed rationality and order. Initiating philosophy and science, Thales believed world mysteries could be discovered in thought. Whereas Thales erroneously proposed that all matter was composed of water as a single physical entity, he boldly and historically broke ranks with traditional beliefs in animism (a supernatural attribution of the soul to plants, animals and inanimate objects) and mysticism to explain natural phenomena.

Anaximander of Miletus who wrote about biology, geography and cosmology proposed a theory that the Universe was boundless and originated from the aperion (the infinite, unlimited or indefinite), rather than from a particular element as water (as Thales held).

Perhaps the first true evolutionist, Anaximander believed that man could not have survived had he always been as he had become. He argued that the long period of care needed by human young precluded non-evolutionary views. Backed by observations of fossil remains, Anaximander contended man evolved from fish at sea.[181]

Ionian philosopher and scientist Pythagoras (circa 500-475 BC) made important contributions to mathematics and geometry which are believed to have later influenced Socrates, Plato, Aristotle and other early Greek thinkers. He is credited with many important discoveries which include the famous Pythagorean Theorem (a fundamental relation in Euclidean geometry), the five regular solid geometries, the Theory of Proportions, the Earth's spherical shape and the identity of the morning and evening stars as the planet Venus.

Of special note, although the Babylonians knew the Pythagorean Theorem at least a thousand years before Pythagoras, it was he who conceived the mathematical proof—that for a right-angled triangle the square of the hypotenuse equals the sum of the squares of the other two sides.

Pythagoras was also devoted to religious mysticism and a believer in the transmigration of souls wherein every soul is immortal and, upon death, enters into a new body.

Greek philosopher Anaxagoras of Clazomenae (a major city in Ionian Asia Minor) was first to give a correct explanation of eclipses and was both famous and notorious for theories, including the claims that the Sun is an incandescent stone (not a god), and that the moon consists of Earthly materials. Yet while disparaging the idea of mythology and anthropomorphic gods, Anaxagoras believed that a transcendent mind set the world in motion, giving it form and order.

Philosophers Leucippus and Democritus developed the idea that the world consisted exclusively of invisible, minute, uncaused and immutable material called atoms. Moving in a boundless void, atoms occasionally collided and combined to produce the visible world. They theorized that this ceaseless movement arose mechanically, not by divine guidance.

Heraclitus and Parmenides posed contrasting thoughts in pre-Socratic Greece where synthesis of opposing views was recognized often to produce synergistic order which was often greater in merit than either extreme.

The philosopher Heraclitus, who lived in Ephesus, a city on the Ionian coast not far from Miletus, offered new meaning to Logos. Opposed to the original meaning—word, speech or thought, he considered Logos divine intelligence or the rational principle governing the Universe. Heraclitus postulated that while things are in constant flux, they are related and ordered by universal Logos.

Heraclitus disbelieved permanent reality. He believed that all things are eternally transient and motile. He reasoned that most humans exist in false dreams and conflict, but its understanding of Logos enables harmonization with deeper realities.

Parmenides from Elea, a Greek city in southern Italy, believed all things consisted of immutable, elementary particles; motion was seen as a sense of illusion. Synthesizing opposing views, atomists drew from Parmenides immutable, elementary particles and from characteristics of ancient thought.[182]

Hippodamus of Miletus is credited as one of the first political theorists who believed that the emergence of the polis, the Greek city-state, and philosophy occurred in concert.

Ancient Greeks considered public life critical to human activity, whereby mythology became dissipated in concert with their belief in the equivalency between principles of reason and politic. Emerging separately from religion, political thought promoted secularism and reason was viewed as the inherent responsibility and right of free men which fundamentally shaped pre-Socratic thinking in all areas of life.[183]

A Sophist movement led by Protagoras in the mid-400s BC believed in independent thinking, that Man is the measure of all things and that each person's opinion, therefore, possessed a level of truth. This came to take on a Machiavellianism emphasis, which as a consequence, defined justice being relinquished to the advantage of the stronger.[184]

Although Protagoras has come to be regarded as a true and sincere pragmatist, later Sophists lost that intellectual honesty. They instead promoted the politically weaponized use of rhetoric—eristic as oppose to dialectic. Eristic is deception by seemingly validating falsehoods.

Sophist students became skilled in the art of eristic, or how to devise plausible arguments supporting untrue claims...today called spin.

The Sophists viewed speculative philosophical theories regarding cosmologies and speculative science implausible and useless. Critics including Aristophanes, Plato and Aristotle later ridiculed this philosophy as dishonest, unethical and shallow, leading to sophistry's derogatory meaning.

The Sophists' relativistic humanism led to broad Athenian skepticism toward all ethical values. Many within the populace found this view lacking and disruptive to social and moral good.

Whereas pre-Socratic natural philosophers had spurned religion and sought material explanations for phenomena, later philosophers—most notably Socrates, Plato and Aristotle—combined mysticism with naturalism. Accordingly, a dichotomy between mysticism and rationalization characterized Greek thought from the time of Socrates (mid-300s BC) on.

During these times of tension, and perhaps to a large degree in reaction to failures of Sophist philosophy, Socrates sought a system of inquiry based upon earnest dialog which was both rational and emotional; a pursuit of knowledge where truth was attainable, yet to be achieved. He also believed that seeking knowledge followed a sacred pathway to God.[185]

Socrates' last dialogue, *Phaedo*, attempted to prove the human soul is immortal. Philosophically most important in this was his description of hypothesis and deduction separate from experiment and observational proofs.[186]

In 399 BC, Socrates, then 70 years old, stood before a jury of 500 of his fellow Athenians accused of refusing to recognize the gods recognized by the state and of corrupting the youth. If found guilty, his penalty could be death. His anti-democratic views had turned powerful authorities and politicians against him, particularly after two of his students, Alcibiades and Critias, had twice briefly overthrown the city government.

After hearing arguments of both Socrates and his accusers, the jury found him guilty by a vote of 280 to 220. Given an opportunity to suggest his own punishment, Socrates might have avoided death by recommending exile. Instead, he initially offered a sarcastic proposal that he be rewarded for his actions. When pressed for a realistic punishment, he next offered to pay a modest fine.

The jury selected a death penalty, and Athenian law prescribed that this be carried out by drinking a cup of poison hemlock. Thus, Socrates became his own executioner, reportedly doing so with equanimity.

Socrates' student Plato (428-347 BC) founded the Academy in Athens, the first institution of higher learning in the Western world. Plato (meaning broad referring to his broad shoulders and forehead) was actually a pseudonym for his real name, Aristocles.

Upon Socrates' execution, Plato lost trust in Athenian democracy—law without standards of justice. Holding to the concept of a supreme ruler, Plato believed that it is essential to establish a standard foundation for philosophies and political systems—one of "absolutism."

Fundamentally a rationalist, Plato's dialogues reflect a dualism between the spiritual and rational. In his *Republic*, he emphasized dialectic and rigorous self-critical logic. In his *Symposium*, Plato championed the supernatural.

Plato proposed geometrical atomism: reality arises from mixtures of space and form. In his view, four elements comprised the physical and biological world: basic particle of fire; basic particle of Earth; basic particle of air; and basic particle of water.

Prominent intellectual Greek thinking at the time viewed the heavens as timeless and unchanging with planets directed by spiritual powers traveling in perfectly circular orbits.

A firm belief in divine, circular order caused Plato to consider it blasphemous to refer to planetary irregularities and their multiple wanderings. He believed critical mathematical reasoning alone would reveal perfect circular orbits of planets, and "the Universe as the living manifestation of divine Reason."

Starkly contradicting Plato, his teacher, Aristotle (384-322

BC), believed that the validity of all philosophical theories must be supported by evidence. Aristotle understood—as others before him had not—that deduction alone is inadequate for achieving reliable knowledge.

Although Aristotle recognized that empirical knowledge is also fallible, ultimate divine truth could be discovered through active intellect which combines empiricism with rationalism.

Aristotle's science and philosophical thinking applied a disciplined and systematic concept of logic aimed at providing a universal process of reasoning that would allow humans to learn every conceivable thing about reality. He characterized deduction as arising from a reasonable argument in which: [W]hen certain things are laid down, something else follows out of necessity in virtue of their being so.

This precept became the basis for what philosophers now refer to as a syllogism, where a logical conclusion is inferred from two or more other premises of very particular forms. This view was at odds with Plato's teachings, which noted deduction simply follows from premises, which lead to seemingly logical conclusions.

At the age of 17, Aristotle had left his birthplace in the small town of Stagira on the northern coast of Greece, to enroll in Plato's Athens Academy. Following Plato's death in 348 BC, his friend Hermias, King of Atarneus in Mysia, invited Aristotle to his court where he met and married the king's niece Pythias.

Seven years later, King Philip of Macedonia handsomely compensated Aristotle to tutor his son, the then 13-year-old Alexander, who became a close friend. Aristotle subsequently became a key and an extremely harsh military counselor during Alexander's Eastern conquest of Persia. He counseled Alexander to be:

> A leader to the Greeks and a despot to the barbarians, to look after the former as after

friends and relatives, and to deal with the latter
as with beasts or plants.

Aristotle's subsequent appointment to head the Royal Academy of Macedonia also involved giving lessons to two other future kings, Ptolemy and Cassander.

Aristotle returned to Athens in 335 BC after Alexander, who succeeded his father, King Philip, conquered the city. Since Plato's Academy was then being headed by Xenocrates, with Alexander's blessings, Aristotle created his own school called the Lyceum. There, he was believed to be the first teacher to organize his lectures into courses and to assign them a place in a syllabus.

Building upon principles worked out by Socrates and Plato, Aristotle developed incipient scientific thought that became a foundation for modern inductive and deductive logic and natural science. Although much of his science was later disproved, he established a basis from which mistakes could be recognized and more reliable knowledge derived.

Aristotle divided the sciences into three general categories: productive, practical and theoretical. The productive sciences not only included engineering and architecture, which yield tangible products such as bridges and houses, but also disciplines such as strategy and rhetoric, where a product is something less concrete, such as victory on the battlefield or in the courts.

Practical sciences, most notably ethics and politics, were those associated with guiding behaviors. Theoretical science involved those that have no product and no practical goal, but in which information and understanding are sought for their own sake. He divided these into three groups: physics (the study of nature), mathematics and theology.

Aristotle's political relationships during those turbulent times brought hard consequences. His relationship with Alexander cooled as the king became more and more megalomaniac, proclaimed himself divine and demanded that Greeks prostrate themselves before him in adoration.

Aristotle was later charged with impropriety for his previous relationship with Alexander's government following the sudden death of Alexander the Great and overthrow of his reign of power in 323 BC. He then fled to Chalcis in 321 BC to escape prosecution and execution and died a year later by drinking a poisonous hemlock beverage. He'd been charged with the same offenses that ended the life of Socrates.

In death, Aristotle left behind an estimated one million words of his surviving works recorded on papyrus scrolls. These scrolls are estimated to represent around one-fifth of his total output. Although his writings generally fell out of use soon afterwards, their retrieval more than seven centuries later significantly influenced Western thought on humanities and social sciences.

Whereas Plato and Aristotle are both ranked by many historians as being the greatest philosophers who have ever lived, it is Aristotle who might better be credited for his contributions to intellectual empiricism, as his work inspired other great thinkers, including Leonardo da Vinci during the Renaissance.

Altogether, the ancient Greeks rapidly achieved a remarkable quantity and intellectual quality of intellectual advancements which did not reappear again until the Renaissance and the Scientific Revolution some 1,400 years later. As Philosopher Bertrand Russell noted:

> *Within the short space of two centuries, the Greeks poured forth in art, literature, science and philosophy an astonishing stream of masterpieces and set the general standards of Western civilization.*[187]

Briefly summarized, a short sampling of these contributions

include:

- Pythagoras, (582-507 BC): recognized that the Earth moves.

- Eudoxus, (408-355 BC): contributed to mathematical astronomy; calculated approximate planetary positions.

- Heraclides, (390-322 BC): proposed that the Earth's rotation causes diurnal movement and that Mercury and Venus always appear close to the Sun because they revolve about the Sun.

- Aristarchus of Samos, (310-230 BC): hypothesized that all planets revolve around a stationary Sun.

- Euclid, (circa 300 BC): made original contributions in natural science; his treatment of elementary plane geometry continues in use today.

- Archimedes, (287-212 BC): a mathematical physicist who contributed many aspects of physics; best known for his principle that a body immersed in a fluid is buoyed up by a force equal to the weight of the displaced fluid.

- Strabo, (63 BC-21 AD): an early geographer who wrote extensively about people and places.

- Galen, (130-200 AD): a physician who showed that arteries carry blood, not air; also advanced knowledge of the brain, nerves, spinal cord and pulse.

- Copernicus, (1473-1543): proved Aristarchus was correct when he developed his heliocentric theory.

A Greek-Roman Era of Turmoil, Division and Recovery

The age of ancient Greek inspiration declined beginning in the 2nd-century with conquest and domination by the Roman Empire. Although this new Roman era began with Grecian defeat in the Battle of Corinth in 146 BC, the Roman Republic had been steadily gaining control over the mainland following its successes in a series of conflicts known as the Macedonian Wars.

Definitive Roman occupation of the Greek world was established following the Battle of Actium (31 BC) in which Augustus defeated Egypt's queen, Cleopatra VII, and then soon afterward, conquered Alexandria, the last great city of Hellenistic Greece.

The Roman era of Greek history continued with Emperor Constantine the Great's adoption of Byzantium as Nava Roma, the capital city of the Roman Empire. In 330 AD, the city was renamed Constantinople, with a general Greek-speaking polity.

Roman domination of lands of conquest adopted a culture of total subordination to Roman Christian orthodoxy with fierce suppression of heretical noncompliance with authoritative doctrine posed by scientific inquiries.

While the famous story of Cleopatra and her lover Marcus Antonius (Mark Antony) immortalized by Shakespeare and other writers may not have monumental historical consequences, it does reveal enormous political turmoil at the time. Born in 69 BC in the city of Alexandria, Egypt, Cleopatra had also been an ally and lover of Julius Caesar until the time of his assassination in Rome in 44 BC.

Caesar's death split Rome into various factions competing for control, the most important of these being armies of Mark Antony and Caesar's former supporter, close friend and later adopted son, Octavian.

In 41 BC, Mark Antony summoned Cleopatra to Tarsus (in

modern southern Turkey). She majestically complied by sailing up the Cydnus River in an opulently decorated barge dressed in robes associated with the Greek goddess Aphrodite. Antony and Cleopatra immediately formed a legendary romantic and political liaison that later ended very badly for both of them.

The union between Cleopatra and Antony provided mutually prized strategic advantages. For Cleopatra it afforded an opportunity to achieve greater power in both Egypt and Rome; for Antony, it represented financial and military support from one of Rome's largest and wealthiest states in his campaign against the mighty Parthians (Parthia was a region in modern northeastern Iran).

The alliance between Cleopatra and Antony was not to be trifled with. She allegedly enlisted his support in successfully executing her half-sister Arsinoe, who had been living in the protection of the Temple of Artemis at Ephesus, to prevent any attempts against Cleopatra's throne. Antony and Cleopatra very publicly traveled to Alexandria together in 40 BC.[188]

Despite an extravagant failure of his Parthian campaign, known today as the Donations of Alexandria, Antony and Cleopatra celebrated a mock Roman Triumph in the streets of Alexandria and issued a proclamation which caused outrage among Octavian's minions in Rome. The declaration (in 34 BC) purported to distribute lands held by Rome and Parthia amongst Cleopatra and her three children fathered by Antony.

To avoid civil war, Antony was not mentioned in the proclamation. Nevertheless, to complicate matters more dangerously, Antony then left Alexandria, traveled to Italy, and, presumably to cool down a rapidly heating conflict with Octavian, married his nemeses' sister, Octavia. This occurred concurrently at the time of his publicly flaunted celebrity relationship with Cleopatra.

Ongoing hostility erupted into civil war in 31 BC as the Roman Senate under Octavian's direction declared war on

Cleopatra and proclaimed Antony a traitor. Later that year, Antony's naval forces were defeated at the Battle of Actium. Octavian then invaded Egypt in 30 BC, laying siege to Alexandria. Hopelessly outnumbered, Antony surrendered, and following Roman tradition, committed suicide by falling on his sword.

Following Antony's death, Cleopatra was taken to Octavian, informed that she would be brought to Rome, paraded through the streets, and following public humiliation, likely executed. According to the ancient historian Plutarch, Cleopatra instead arranged to have a poisonous asp (an Egyptian cobra) brought to her concealed in a basket of figs and intentionally died from its bite along with two female servants. (Other historians, including Joyce Tyldesley, believe that Cleopatra used either a poisonous ointment or vial to commit suicide.)[189]

In any case, the deaths of Antony and Cleopatra (the last monarch of the Ptolemaic Empire) left Octavian the undisputed master of the Roman world.

In 330 AD, the Roman emperor Constantine dedicated a New Rome (which he later named Constantinople) at the site of the ancient Greek colony of Byzantium located on the European side of the Bosporus linking the Black Sea to the Mediterranean—an ideal trade point between Europe and Asia. Five years later, at the Council of Nicaea, Constantine established Christianity as Rome's official religion.

Being monotheistic, Christianity undermined religious traditions of the Roman state which regarded the emperor as a god. This, in turn, tended to weaken the emperor's public authority and credibility.

In 364, following Constantine's death in 337, Emperor Valentinian I divided the Roman empire into western and eastern sections, putting himself in power in the West, and his brother, Valens, in the East. The western part of the empire spoke Latin and was Roman Catholic, while the eastern part spoke Greek and worshiped under the Eastern Orthodox branch of the Christian church.

It is important to note here that references to the subsequent fall of Rome relate only to the western part of the empire. The eastern half of the empire, the Byzantine Empire, continued to exist for centuries.

Issues other than religious disputes contributed to the fall of the western part of the original Roman Empire. A decrease in agricultural production led to higher food prices leading to a large trade deficit with the eastern half. Whereas the west continued to desire eastern luxury goods, they had nothing to exchange.

To make up for a lack of money, the western government began to produce coins with less silver content which led to inflation. Adding to this problem, piracy and attacks from Germanic tribes disrupted the flow of trade with particular pain to the west.

Wave after wave of Germanic barbarian tribes, including Visigoths, Vandals, Angles, Saxons, Franks, Ostrogoths and Lombards, swept through and ravaged the Roman Empire. Romulus Augustulus, the last of the western Roman emperors, was overthrown by the Germanic leader Odoacer in 460 AD. Odoacer became the first Barbarian to rule Rome.

The order that the Roman Empire had brought to Western Europe over 1,000 years had ended.

Meanwhile, the Eastern Roman Empire (variously known as the Byzantine Empire, or Byzantium) tended to flourish. Although ruled by Roman law and political institutions with its official language as Latin, Greek was also widely spoken, and Greek history, literature and culture were included in education systems.

In 451, the Council of Chalcedon officially established a division of the Christian world into five patriarchates, each ruled by a patriarch: Rome (where the patriarch would later be called pope), Alexandria, Antioch, Jerusalem and Constantinople (where the patriarch emperor would head both church and state).

Even after the Islamic Empire later absorbed Alexandria, Antioch and Jerusalem in the seventh century, the Byzantine emperor would remain leader of most eastern Christians.

Justinian I took power over the Byzantine Empire in 527 and controlled most of the land surrounding the Mediterranean Sea, including part of the former Western Empire (such as North Africa) captured by his armies. Justinian established a Byzantine legal code that endured for centuries, and built great monuments including the spectacular domed Hagia Sophia (Church of Holy Wisdom).

By the time of Justinian's death in 565, the Byzantine Empire reigned as the largest and most powerful state in Europe. At the same time, however, debts incurred in wars had left the empire in dire financial straits. As a result, Justinian's successors were forced to levy heavy taxes upon Byzantine citizens to keep the empire afloat.

Making matters worse, Justinian's vain struggles to maintain control of his territories had left the imperial army stretched very thin. Seventh and eighth-century attacks from the Persian Empire and from Slavs combined with internal political instability and economic regression threatened the empire's survival.

A new and more ominous threat arose in the form of Islam, a religion founded by the prophet Muhammad in Mecca in 622. Muslim armies soon began to attack the Byzantine Empire by storming into Syria.

Eighth and ninth century Byzantine emperors (beginning with Leo III in 730) began to deny homage of holiness to all icons of non-Christian worship. Referred to as Iconoclasm—literally the smashing of images—the movement waxed and waned under various leaders until 843 when a Church council under Emperor

Michael III ruled in favor of religious displays.

The late 10th and 11th centuries enjoyed a golden age of Macedonian dynasty art and culture initiated by Michael III's successor, Emperor Basil. Although his Byzantium stretched over less territory, it controlled more trade, wealth and prestige than under Justinian.

Macedonian rulers began restoring churches, palaces and other cultural institutions. They also promoted the study of ancient Greek history and literature, establishing Greek as the official state language.

Monks administered many institutions (orphanages, schools and hospitals) in everyday life, and missionaries won many Christian converts among Slavic peoples of the central and eastern Balkans (including Bulgaria and Serbia) and Russia.

Killing for Love of God and Neighbor

The golden age ended when the late 11th century witnessed the beginning of a series of holy wars waged by European Christians against Near Eastern Muslims between 1095 and 1291. Termed the Crusades, the conflicts began when Constantinople Emperor Alexius I turned to the Western Empire for help to defend against attacks from Central Asian Seljuk Turks.

The Crusades also arose partly as a consequence of overpopulated warrior class in Europe, causing local warfare. Population growth led to unrest and soil infertility. This unrest became aggravated by the requirement that younger sons were rejected from land that only eldest sons inherited.[190]

The Church promoted taking the cross as a demonstration of love of God and love of neighbor. Likewise, knights were taught that to be a good Christian knight required undertaking acts of such love and charity in God's name.

Commenting on Muslim victories in the holy land, French monk Bernard of Clairvaux wrote:

If we harden our hearts and pay little attention…where is our love of God, where is our love for our neighbor?

To maintain a hold over warrior aristocracy, a papacy fearful of losing power directed warfare. As a huge incentive, the Church taught at that time that an individual's sins could be remedied, at least in theory, by acts of penance that demonstrated remorse and a desire for forgiveness. As communications through Central Europe improved, and Italian trade in the Mediterranean increased, more Western European people than ever before could journey or make a pilgrimage to the Holy Land and seek penance for past sins.

Pope Urban II responded to Emperor Alexius' call for armed assistance from Turk invasions by declaring a holy war which commenced the First Crusade (1096-1102). As a result, the Europeans captured Jerusalem in 1099. The Muslims, however, quickly unified against the Christian invading and occupying force, and the two groups battled in subsequent wars for control of the Holy Land. By 1291, the Muslims firmly controlled Jerusalem and the coastal areas, which remained in Islamic hands until the twentieth century.[191]

Christians were promised that the Crusades offered a path to salvation for those who participated. As the 12th-century French monk Guilbert of Nogent instructed:

God has instituted in our time holy wars, so that the order of knights and the crowd running in its wake…might find a new way of gaining salvation. And so they are not forced to abandon secular affairs completely by choosing the monastic life or any religious profession, as used to be the custom, but can attain in some measure God's grace while pursuing their own careers, with the liberty and in the dress to which they are accustomed.

Those who took up the Cross were to become recipients of both spiritual and Earthly rewards, in addition to gaining indulgences—forgiveness of sins—they were also to receive shares of plunder from conquests, forgiveness of debts, freedom from taxes, fame and privileged access to political influence.

As a result, millions of people, Christians and non-Christians, soldiers and noncombatants lost their lives. In addition, the debt incurred along with other economic costs associated with multiple Middle East excursions impacted all levels of society—from individual families—to the many villages and cities that were destroyed in the Crusaders' wake.

Altogether, Christians made eight unsuccessful attempts, including an unsanctioned Children's' Crusade, to recapture the Holy Sepulcher, the Holy Grail and the Holy Land from the Muslims.

In his book *The History of the Rise and Fall of the Roman Empire,* eighteenth-century writer Edward Gibbons refers to these Crusades as events in which...

> *...the lives and labours of millions, which were buried in the East, would have been more profitably employed in the improvement of their native country.*[192]

Approximately beginning with the reign of Emperor Michael VIII in 1261, the economy of the once-mighty Byzantine state was crippled, never to regain its former stature.

Emperor John V Palaiologos, the son of Emperor Andronikos III, oversaw the gradual dissolution of imperial power amid numerous civil wars and witnessed continuing ascendancy of the Ottoman Turks.

In 1367, John V appealed to Pope Urban V in the West for help to stave off the Turkish threat and to end a schism between the Byzantine and Latin churches by voluntarily submitting the patriarchate to the supremacy of Rome. In 1369, he traveled through Naples to Rome and formally converted to Catholicism in St. Peter's Basilica where he recognized the pope as the supreme head of the Church. This concession was not enough, however, to end the split.

Impoverished by wars, John was detained as a debtor in Venice on his return trip from Rome and was later captured on his way back through Bulgarian territories. In 1371, he submitted to the authority of Ottoman sultan Murad I.[193]

As a Turkish vassal state, Byzantium paid tribute to the sultan and provided him with military support. Although the Byzantine Empire gained sporadic relief from Ottoman oppression under John's successors, the rise of Murad II as sultan in 1421 marked the beginning of the ending of a final respite.

Murad revoked all Byzantine privileges and laid siege to Constantinople. Murad's successor, Mehmed II, launched the final turnover of Constantinople to Turkish dominance when he triumphantly entered the Hagia Sophia and converted it to the city's leading mosque.

The fall of Constantinople in 1453 marked the end of a once-glorious Byzantine Empire, ushering in the long reign of the Ottoman Empire.

The centuries leading up to the final Ottoman conquest left a rich Byzantine legacy of philosophy, literature, art, science and invention which had flourished even as the empire itself had faltered.

Long after its end, Byzantine culture and civilization continued to exercise an influence upon countries that practiced its Orthodox religion, including Russia, Romania, Bulgaria, Serbia, Greece and others.

This broad and deep heritage came to exert particularly

Larry Bell

great influence on Western intellectual thought and traditions as Italian Renaissance scholars sought help from Byzantine scholars who had fled Constantinople for Italy in translating Greek and Christian writings.

The Clouded Dawn of Renaissance

A Renaissance period dawn which began in 14th century Italy witnessed the emergence of numerous freethinkers, humanists and scientists who openly declared opinions opposed to religious ecclesiastical sanction. In parallel, that propensity for opposition to liturgical doctrine ended very badly for some of them as well as zealous tribunals exacted harsh penalties for suspected heresy.

Originally seen as a pathway to the transcendental world of God, free-thinking scholastic rationalism which set the stage for this Renaissance had begun to upset Church dogma centuries earlier.

Transitional to the Renaissance, the Roman Church had become rife with corruption. By the 13th century, simony, the selling benefices to the highest bidder, was a common practice which heaped huge revenues into Church hierarchy coffers.

A political vacuum left by Imperial Rome transferred great power to popes who enacted doctrines which were anything but Christ-like. A long era of deep Church corruption and cruel heresy purges institutionalized torture, mutilation and incineration which continued to into the 17th century High Renaissance:

- Pope Gregory VII (1020-1085) declared that "foul plague of carnal contagion" was responsible for clergy venality and authorized the laity to withdraw support from priests who did not renounce their wives and children. Abandoned by their husbands, church wives and children became destitute and starving. Under Gregory, simony became even more pronounced.

182

Gregory also curtailed intellectual freedom, insisting that education was solely the Church's responsibility.

- Pope Innocent III (1198-1216) exacted torture for women and freethinkers in Southern France.

- In 1233, Pope Gregory IX formed a formal Catholic Church inquisition tribunal which sanctioned torture to obtain heretical confessions. In 1229, the Inquisition set up courts in Toulouse, rounded up many women accused of heresy, racked them until confession and burned them publicly.

- In 1478, Spain's King Ferdinand and Queen Isabella appointed a cruel Dominican, Friar Torquemada, as Chief Inquisitor. He sanctioned torture of girls as young as 13, then sentenced them to die by burning.

- In 1542, Paul III assigned the Inquisition to the Holy Office. Pope Paul IV who followed (1555-1559) terrified Italians by establishing severe religious persecution, removed nude paintings and statues, painted over Michelangelo's indiscretions in the Sistine Chapel and burned thousands of books.

- Galileo was famously placed under house arrest and brought before the Inquisition in 1630 for arguing against the geocentric model concept of Ptolemy backed by the Roman Catholic Church that the Sun orbited around the Earth (and not the other way around). Galileo's works advocating Copernicanism were then banned, and a papal order prohibited him from teaching, defending or even discussing his discovery.

- In Germany, Kepler's works were also banned by the papal order.

Whereas astronomy gained momentum during the 12th century

through assimilation of Greek and Islamic knowledge, Church-sanctioned scholars continued to believe that the heavens are spheres with circular motion, with Earth as the central body. This Earth-centrality could not be adjusted to correlate with the cycles, epicycles and eccentric of Ptolemaic astronomy.

Scientific inquiry was regarded as a curse by those who feared challenges it posed to Church doctrinaire authority. For example:

- Medical practices based upon meticulous observation demanded by ancient physician Greek Hippocrates had become replaced with trust in miraculous healing which employed prayer, chants, potions, horoscopes and amulets. Disease was explained as the divine punishment for sin.[194]

- The Paris medical faculty proclaimed that a disastrous siege of Black Death (bubonic plague) between 1347 and 1351 was caused by "corruption of the air" following from the conjunction of Jupiter, Saturn and Mars.

- The anatomical study of cadavers (other than those of criminals following execution) was forbidden by the Church until the 14th century.[195]

Church glorification by building beautiful, artistic structures such as St. Peter's Basilica was financed by sales of spiritual indulgences—payments for the remission of punishment for sins which also funded the Crusades. This corrupt practice led the German Augustinian monk Martin Luther (1483-1546) to rebel and break away from the Catholic hierarchy.

Luther's Protestant Reformation movement initiated the return to a fundamentalist Old Testament biblical emphasis which divided European Christianity into Catholic and Protestant followers.

Beginning in the late 1400s, violent persecution erupted

between Catholics and Protestants, and also among Protestant sects. Religious wars waged in the name of God ravaged on in Europe for 150 years.

Thirteenth century Dominican Albertus Magnus (Albert the Great) was among the first medieval thinkers to distinguish between theological knowledge and science. Albert's broad interests included prolific writings about botany, astronomy, chemistry, physics, biology, logic, metaphysics, meteorology and zoology. His premise that faith and reason are not incompatible sources of knowledge inspired the major work of his most famous pupil, his Dominican colleague and friend, Thomas Aquinas.[196]

Thomas Aquinas (1225-1274) fully amalgamated Aristotelian physics with Church doctrine. Like Albertus Magnus, he believed that the combined thought forms, faith and rationalism, benefited the Christian cause by arousing deeper understandings to support beliefs.

Also like Magnus, Aquinas was an intellectual who merged Greek with Christian ideals into a single holistic doctrine called Summa Theologica. Aquinas agreed with Socrates—that evil stems from ignorance, less from malice—and his version of Aristotle's work came to be established as a dominant philosophy.

The Church canonized Aquinas scholar-saint in the mid-13th century.[197]

Whereas Franciscan scholars had argued for a sharp distinction between natural philosophy and theology, in their success, rational inquiry severed from theology resulted in a great boon to science. Relieved of religious restraints, free inquiry allowed those of faith to do undertake unfettered scientific research.

Franciscan scholar Roger Bacon (1220-1292) who is celebrated as the first true experimental scientist in Britain, wrote in his *Opus maius* (Experimental Science):

Having laid down fundamental principles of the wisdom of the Latins so far as they are found in language, mathematics, and optics, I now wish to unfold the principles of experimental science, since without experiment nothing can be sufficiently known.

Imprisoned for 15 years, Bacon's ideas were viewed unfavorably by the Church—empirical emphasis over metaphysical speculation was viewed his greatest sin.[198]

An acceptance of scholasticism gradually arose with Roman Catholic Church doctrine which held that God and religion could be justified by rationalism, a contention that natural knowledge offers a legitimate path to religious contemplation and mystical ecstasy.

British philosopher and priest William of Ockham emphasized the attainment of knowledge through logical thinking. Ockham established the philosophical principle that "entities are not to be multiplied beyond necessity." Known and still popular today as Ockham's razor, it teaches that until proven otherwise, the preferred choice of answer to an uncertainty is the simplest explanation that best fits the facts.[199]

Intellectual growth inspired by scholasticism, along with reinstatement of Greek knowledge, established a springboard for great scientific and creative achievements which are associated with Renaissance genius.

As explained by Richard Tarnas:

The medieval gestation of European culture had approached a critical threshold, beyond which it would no longer be containable by old structures.

Nevertheless, the community of scholars remained to be

polarized between opposing penchants of faith-revelation and evidence-logic.[200]

The free-thinking Renaissance which emerged in the 14th peaked in the 15th and 16th centuries and ended in the 17th century as zealous religious police reinstated witch hunts for unbelievers. Luther's religious conservatism Reformation movement again separated theology from science whereby faith and reason clashed anew.

Free thinking attached to 16th and 17th century scientific and technological advancements provoked religious backlashes. Nicolaus Copernicus's heliocentric astronomical model published in 1543 which placed Earth and other planets along circular paths around a motionless Sun disrupted the concept of a human-centric, therefore church-centric, Universe.

Johannes Gutenberg's 15th-century invention of movable type printing along with innovations in rapidly casting type in hand molds dispersed ideas regarded to be intimidating to religious authorities beyond their control. Such backlash provoked renewed suppression of free-thinkers, which in turn, stimulated even greater intellectual curiosity and determination.

Italian philosopher and mystic Giordano Bruno (1548-1600) was burned at the stake by the Inquisition, not as a martyr for science, but as a defender of his own unique philosophy of mysticism. Accused of diabolic practices, it appears that his unforgivable sin was to deny Christ's divinity.

Sparks of Immortal Genius

Above all, the Renaissance is remembered as a period of creative genius, an era characterized by humanist literary and artistic expression immortalized by Petrarch, Boccaccio, Bruni, Alberti, More, Machiavelli and Montaigne in the early period—later by masterworks of Shakespeare, Cervantes, Michelangelo, Raphael and Leonardo da Vinci.

Larry Bell

Among all of these creative giants, Leonardo da Vinci epitomizes the Renaissance ideal of a Universal genius. Still widely recognized as one of the greatest artists of all time, he also engaged and excelled in an amazing variety of engineering and scientific endeavors ranging from machines for military defense and warfare, architecture and cartographical mapping, innovative concepts for flight and detailed studies of human anatomy and botany.

Leonardo's remarkable diversity recognized none of today's prevalent boundaries and mutually-exclusive polarities between science and art. His successes in both types of endeavors drew upon broad curiosity about the natural world which were combined with highly developed observational skills and dedicated attention to technical details.

Unlike Michelangelo, Raphael and most other religious artists of his time, Leonardo saw the world as logical rather than mysterious. He was also a fundamentally different kind of scientist than Galileo, Newton and others who followed him. Although Leonardo's observational approach to science attempted to understand a phenomenon by describing and depicting it in utmost detail, it did not emphasize experiments or theoretical explanations.

Consequently, while many of Leonardo's concepts either weren't technically feasible due to lack of scientific information, others were so ahead of his time that they required modern metallurgy and engineering advancements which only occurred centuries later. Examples of both include a fundamentally infeasible flapping ornithopter flying machine, and another with a helical rotor. His concept for armored double-hull ships used for military and commercial applications was successfully adopted long after his death.

At the same time, some of da Vinci's numerous inventions were not only implemented but were highly beneficial in his day. His designs for an automated bobbin winder, and a machine for testing the tensile strength of wire, proved revolutionary even then.[201]

Leonardo was born out of wedlock (1452) in the Vinci region of the Medici-ruled Republic of Florence, home to many great artists and philosophers. Luminary figures included painters Piero della Francesca and Filippo Lippi, sculptor Luca della Robbia and writer and architect Leon Battista Alberti.

Such influential leaders were followed by Leonardo's teacher, the renowned artist Andrea del Verrocchio, Antonio del Pollaiuolo and painter-sculptor Mino da Fiesole. Leonardo was a contemporary of Botticelli, Domenico Ghirlandaio and Perugino...and most notably, Michelangelo and Raphael.

While all three of these High Renaissance masters were contemporaries of one another, they were not actually of the same generation. Leonardo was twenty-three years old when Michelangelo was born and was thirty-one when Raphael was born. Raphael died at age 37 in 1520...the year after Leonardo died. Michelangelo then lived on for another 45 years.[202]

Leonardo was first introduced to informal studies of Latin, geometry, mathematics and many other subjects as 14-year-old an apprentice in Verrocchio's studio. There, he, along with other interns, was exposed to drafting, chemistry, metallurgy, metal working, plaster casting, leather working and carpentry, as well as drawing, painting, sculpting and modeling. By 1472, at age 20, Leonardo was considered as a master in the Guild of Saint Luke, a respected organization of artists and doctors of medicine.[203]

Leonardo conducted morgue studies which carefully recorded physiological effects of age and human emotions—rage in particular. He also dissected and studied anatomies of various animals including horses, cows, monkeys, bears, birds and frogs. Comparing their anatomies with humans, he concluded that the heart is central to the circulatory system in all. He demonstrated how the heart functions by creating wax models of the cerebral ventricles and a glass aorta to observe blood circulation through the aortic valve using grass seeds in water to watch flow patterns.

As with other great Italian painters of his time, Leonardo's works reflected a fresh, new natural world perception—fidelity, vigor, graceful human body representation and mastery of perspective. In addition to such masterpiece works as the *Mona Lisa, The Last Supper,* and his iconic drawing of the *Vitruvian Man,* many of the more than 13,000 pages of recorded drawings, notes and scientific thoughts on nature he produced throughout his life fuse art with natural philosophy—the forerunner of modern science.

In 1500, Leonardo, along with his assistant and a mathematician friend, fled Milan for Venice as guests of monks in the Monastery of Santissima Annunziata and St. John the Baptist. In 1502 he traveled throughout Italy in the service of Cesare Borgia, the son of Pope Alexander VI, as his chief military architect and engineer. In this role, he created a strategic map which described defensive methods to protect Cesare Borgia's stronghold town Imola from naval attacks. Leonardo produced another defense map for the Chiana Valley which included a proposed sea dam to supply Florence with water throughout all seasons.

Leonardo went to Florence in 1507 to sort out some estate problems with his brothers following the death of their father, then moved back to Milan in 1508. Between 1513 and 1516, he spent extended periods in the Vatican in Rome under Pope Leo X where Raphael and Michelangelo were also very active.

By this time, Leonardo had cultivated important royal and religious mentors. King Francis I of France recaptured Milan in 1515, and Leonardo was known to be present at a meeting of Francis and Pope Leo X in Bologna. Francis commissioned him to create a mechanical lion that could walk forward and then open its chest to reveal a cluster of lilies. In reward for his services, Francis awarded Leonardo a comfortable pension and prestigious manor house, now a public museum, near the king's residence at the royal Château d'Ambroise.

As historian Liana Bortolon wrote in 1967:

> *Because of the multiplicity of interests that spurred him to pursue every field of knowledge...Leonardo da Vinci can be considered quite rightly, to have been the universal genius par excellence, and with all disquieting overtones inherent in that term. Man is uncomfortable today, faced with a genius, as he was in the 16th Century, five centuries have passed, yet we still view Leonardo with awe.*[204]

Larry Bell

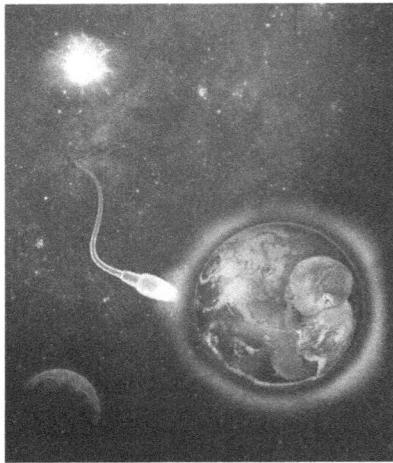

Part Eight: Revolutions of Science, Culture and Industry

THE 16TH THROUGH 19TH centuries marked trending towards independent thinking that provided revolutionary scientific, technological and cultural foundations which are now generally taken for granted in contemporary society.

Much of the impetus for this new and irreversible era of proactive inquiry, free-thinking and innovation, originated to a large extent in 16th century Western Europe during a Scientific Revolution which blended into an 18th century period of philosophical Enlightenment. These developments, in turn, provided knowledge, methodologies and inventions that gave rise to a globally-transformational Industrial Revolution.

Early free-thinkers whose "heretical" scientific inquiries and concepts challenged Roman Catholic Church and Protestant doctrines famously did so at great peril. This oppressive circumstance gradually diminished in combination with the disintegration of the Holy Roman Empire. This disintegration was influenced by growing reactions against Church corruption and a century of crisis among divergent Christian beliefs that weakened intellectual conscriptions to religious orthodoxy.

A thirty-year religious war between Catholic and Protestant states (1618-1648) irrevocably changed the map of Europe.

The complex saga began as a battle between Catholic and Protestant states when future Holy Roman emperor Ferdinand II in his role as king of Bohemia attempted to impose Roman Catholic absolutism on his domains along with Austria. After Ferdinand triumphed in a five-year struggle, King Christian IV of Denmark saw an opportunity to gain valuable territory in Germany to balance his earlier loss of Baltic provinces to Sweden.

Christian's ultimate defeat in 1629 finished Denmark as a European power. After ending a four-year war with Poland, Sweden's Gustav II Adolf invaded Germany, winning over many of its princes to his anti-Roman Catholic, anti-imperial cause.

Poland, in turn, having been drawn in as a Baltic power coveted by Sweden, pushed its own ambitions by attacking Russia and establishing a dictatorship in Moscow under its future king Wladyslaw IV Vasa. A Russo-Polish Peace of Polyanov in 1634 ended the Poland tsarist throne, but freed Poland to resume hostilities against its Baltic archenemy. Sweden had become embroiled in German conflicts where three denominations vied for dominance: Roman Catholicism, Lutheranism and Calvinism.

Meanwhile, a network of Protestant town and principalities that relied on anti-Catholic powers of Sweden and the United Netherlands threw off the yoke of control by Spain following an 80-year-long struggle. A parallel struggle involving the French rivalry with the Hapsburg Empire of Spain ended with France as the victor.

As the Thirty Year War evolved, it became less about religion and more about which groups would ultimately govern Europe.

By 1648, the balance of power in Europe had dramatically

changed. Spain had lost control not only the Netherlands, but also its dominance in Western Europe. France became a chief Western power, while Sweden controlled the Baltic. The United Netherlands became recognized as an independent republic, and member states of the Holy Roman Empire were granted full sovereignty.

The ancient notion of a Roman Catholic Empire of Europe headed spiritually by a pope as emperor was finally abandoned.

A Scientific Revolution

The aftermath of the Thirty Years War more fully catalyzed a new scientific era which began a century earlier in which members of the educated class turned increasingly to direct observations and methodological empirical evidence to investigate natural phenomena.

A crowning achievement of this centuries-long period was great strides in better understanding natural laws which govern the dynamic architecture of our planet, Solar System and celestial neighborhood.[205]

Many historians are inclined to mark the Scientific Revolution threshold beginning with Nicolaus Copernicus's heliocentric hypothesis which he published in 1543. His theory, one which influenced other important visionaries who followed, positioned the Sun near the center of the Universe, motionless, with the Earth and other planets orbiting around it in circular paths modified by epicycles and at uniform speeds.[206]

Soon after Copernicus's proposition confounded Church teachings of an Earth-centered Universe, Danish astronomer Tycho Brahe (1546-1601) observed that with the exception of the Earth, all five planets, known at that time, revolved around the Sun. By holding to an Earth-centric belief, he evaded Church wrath.[207]

William Gilbert (1544-1603) undertook original studies

which determined that our planet Earth somehow causes all magnetic phenomena in its environment. Applying experiments with observable proofs, Gilbert also distinguished differences between electricity and magnetism.

Johannes Kepler (1571-1630) validated Copernicus's radical heliocentric theory and postulated that planetary orbits are actually elliptical. Kepler recognized that the Sun controls planetary motion, and determined that the orbital periods of planets relate to their distances from the Sun.

Also following Gilbert's lead, Kepler explored the possibility that there might be some kind of a magnetic cosmic force that controls planetary positions and movements—the first known insight into gravitational theory. Kepler's empirical data and mathematics provided an essentially correct planetary picture which formed a fundamental basis for Isaac Newton's later seminal advances in cosmic science.

The same year that Copernicus published his planetary motion laws, Galileo Galilei (1564-1632) constructed a telescope to view planets and moons far better than ever before. Galileo discovered that Jupiter and its four moons move in concert around the Sun, and that Earth and its Moon revolve around the Sun as well. Observing phases of Venus, Galileo concluded that it also revolves around the Sun.

Often regarded as the Father of Modern Physics, Galileo's 1610 *The Starry Messenger* publication changed the Universe structure concept beyond conventional imagination, one where even the Milky Way galaxy seemed to be part of a much larger Universe where stars extended outward in every direction.

In addition to supporting Copernican theory, Galileo showed that its principles could be characterized by mathematical formulations. For example, he demonstrated the inseparability of time and motion by a discovery that distances of fallen objects accelerate as the square of time. A body falls 16 feet in one second; therefore, it will fall 4 x 6 = 64 feet in two seconds. This proven formula refuted Aristotle's then-accepted common-sense contention that fall rate is proportional to weight.

While the pope initially showed some enthusiasm towards Galileo's cosmological discoveries, his ideas later sparked vitriolic criticisms with powerful Church and university critics. Frenzied fear of losing control over the faithful, his most dangerous notion that science took precedence over theology, resulted in being labeled a heretic which drew attention from the Inquisition tribunal.

One charge levied against Galileo was that he was unable to present quantitative evidence to support his radical heliocentric-rotating-Earth cosmology theory. English mathematician, physicist Sir Isaac Newton's (1642-1727), accomplished this proof in his 1687 book *Principia* which provided equations for gravity and for accurately predicting planetary motion. After determining Earth's orbital size, distances to nearby cosmic bodies could be calculated. Triangulation and stellar parallax then enabled mathematical distance determination.

Nevertheless, the Church snubbed Newton's clear proof of Galileo's correctness and held to their erroneous beliefs for five more centuries.[208]

Although Cardinal Bellarmine banned Galileo's writings, they were smuggled north where the Western intellectual struggle resumed. Ironically, in 1930, Pope Pius XI canonized Bellarmine.[209]

Observing that polished blocks move across a polished table more readily than on unpolished surfaces, Galileo predicted that movement of objects on frictionless surfaces proceeds forever. This concept of inertia was advanced by René Descartes at a very small scale, where: a particle at rest remains at rest without an impetus or push, while a particle in motion moves in a straight line at the same speed unless deflected by collision with another particle.

Descartes (1596-1650) perceived the Universe as a machine

which is ordered by mathematical laws, an atomist concept wherein an infinite number of particles moved as imposed by God at their creation. He pondered that some unknown invisible force must hold planets in their orbits.

A strong Catholic, Descartes championed the concept of a scientific-religious dualism—wherein human spirituality and mechanical Earthly matter coexisted separately.

Descartes' reductionist philosophy advocated that the natural physical world can be best understood when analyzed from unique arrangements of its simplest parts. He believed that science embodied use of hypothesis and experiment to clearly characterize the nature of all things in terms of extension, size, shape, number, duration, specific gravity and relative position.[210]

In *Discourse on Method*, Descartes famously concluded his own spiritual existence could not be doubted. He originally wrote: *je pense, donc je suis* (French), *cogito, ergo-sum* (translated into Latin), and later appearing in English, *I think, therefore I am.*

Late 17th century scientist Robert Hooke (1635-1703), who is most broadly recognized for his micrographic discovery and naming cells, is less known for his provident theory that the same force governing planetary motion also applies to falling bodies.

Hooke's writings dating back to 1664 propose a concept of gravitational attraction which he describes as:

> …*such a Power, as causes Bodies of a similar or homogeneous nature to be moved one towards the other, till they are united…Planets are of the same nature as the sun and hence are attracted to it. Comets are not related, and they are repelled.*[211]

Although Hooke made no direct reference to a centrifugal force providing gravity, he recognized that a body revolving in orbit must be continually diverted from its inertial path by some force directed toward a center.

Hooke also stated his conviction that gravity decreases in power in proportion to the square of the distance, a view that may very well have influenced Newton who himself acknowledged in 1686 that correspondence with Hooke had stimulated him to demonstrate that an elliptical orbit around a central attracting body placed at one focus entails an inverse square force.

While having proposed the problem of the dynamics of elliptical orbits, Hooke also admitted his inability to solve it. That challenge was successfully addressed by Newton.

Isaac Newton's work, which synthesized Descartes' mechanistic philosophy, Kepler's laws of planetary motion and Galileo's laws of terrestrial motion explained this force. Reasoning that an apple falling from a tree demonstrated a force counteracting centrifugal force, Newton's mathematical analysis of this force, gravity, explained a great previous mystery.

Newton concluded that the Sun pulls planets with an attractive force that decreases inversely as the square of the distance from the Sun, and bodies falling toward the Earth conform to the same law. Newton also calculated elliptical planetary orbits and their speeds.

Newton's contributions to science are remarkable. He invented calculus before age 23. By 1684, he determined the mass and distance laws of gravity. By 1687, he formulated the three laws of motion and developed a method of quantifying centrifugal force.

Newton also determined that prisms break white light into component colors—different light refraction for different wavelengths. Then recognizing that chromatic aberrations in microscopes and telescopes relate to light refraction, the bending of light waves, he illustrated this phenomenon using a prism and then built a reflecting telescope corrected for chromatic aberration.[212]

Whereas Newton and Aristotle believed that light travels instantaneously, in 1676 Danish astronomer Ole Romer showed that light has a finite speed. Observing that eclipses of Jupiter's several moons showed different timing than predicted depending upon variable distances separating Jupiter from Earth, Romer ascribed these discrepancies in light travel times.

Other great thinkers of the time were directing their attention to a variety of different scientific inquiries. Among these, the invention of the microscope opened up new worlds of understanding at a very tiny scale.

During the early 1600s, Galileo reportedly described an ability to focus his telescope to view small objects close up by looking through the "wrong" end. He subsequently improved the design of the device in 1624 based upon a compound instrument with a convex objective and a convex eyepiece (a Keplerian microscope) he had seen in London, perhaps one designed by Cornelius Drebble.

Dutch natural history student Anthony van Leeuwenhoek applied the microscope invention in 1650 to discover protozoa, rotifer, red blood corpuscles, capillaries and sperm cells in semen. The latter observation that a sperm combines with a female egg represented a radically new concept of reproduction which dispelled a mystical notion of spontaneous life generation.

In 1795, about a century and one-half later, English physician Edward Jenner invented the smallpox vaccination to save many lives from ravaging outbreaks of microscopic pestilence.[213]

A Period of Enlightenment

Dated generally between 1685 and 1815, the eighteenth century Age of Reason or Enlightenment was a movement that spread throughout Western Europe, England and the American colonies which dominated and radically reoriented intellectual, philosophical and political discourse.

Following closely on the heels of the Renaissance, The Enlightenment was, at its center, a celebration of ideas about what the human mind was capable of and what could be achieved through deliberate action and scientific methodology to advance individual opportunity and progress.

A strong new spirit of egalitarianism held the purpose of fair treatment of all people. Contending that freedom and democracy were fundamental rights of all people, not gifts bestowed upon them by beneficent monarchs or popes, growing numbers of voices openly expressed sharp criticism of the Church oppression and obstruction of free inquiry.

The Enlightenment championed scientific and humanistic knowledge, a quest for secular understanding to further human rights, dignity and progress. Philosopher Bertrand Russell characterized the movement as a revaluation of independent intellectual activity—a clearing of religious thought to allow seeking of knowledge where darkness prevailed.[214]

French philosopher Denis Diderot (1713-1784) produced the signature publication of the period, *Encyclopedie*, which brought together leading authors to produce an ambitious compilation of knowledge. Diderot also wrote many of the articles himself, a strategy designed, in part, to avoid French censors.

Diderot's monumental compendium of then-current knowledge challenged Roman Catholic Church authority and also that the aristocratic French government, both of whom tried unsuccessfully to suppress it.

Diderot championed the value and uniqueness of the individual and promoted an optimistic belief that all knowledge could be acquired through scientific experimentation and the exercise of reason. He also advocated that education should be tailored to the abilities and interests of the individual student, and that students should learn to experiment and conduct research rather than simply acquire outside knowledge.

German philosopher Immanuel Kant (1724-1804) worried that accepted knowledge had grown too dependent upon the thinking of just a few people and coined the Age of Reason motto: *Dare to know! Have courage to use your own reason!* [215]

Above all, Kant insisted that every rational being had both an innate right to freedom and a duty to enter into a civil condition governed by a social contract in order to realize and preserve that freedom.

Kant argued that the power of the state must be limited to protect citizens from the arbitrary exercise of authority, wherein the concept of "state" can be variously translated as the legal state, state of rights, or constitutional state in which the exercise of government power is constrained by law. This approach is based upon the supremacy of a country's written constitution—a supremacy which must create guarantees of a peaceful life as a basic condition for the happiness of its people and their prosperity.

Influential English philosopher John Locke (1632-1704) argued that human nature was mutable, and that knowledge must be gained through accumulated experience rather than by accessing some sort of outside truth.

Locke championed a principle that all people are equal and independent, with a natural right to defend their "life, health, liberty, or possessions."

Together with essays on religion, which provided an early model for the separation of church and state, Locke deeply influenced America's founding documents. Thomas Jefferson echoed Locke's concepts in the first sentence of the Declaration of Independence: "Human equality, and the right to life, liberty, and the pursuit of happiness." [216]

Jean-Jacques Rousseau and Voltaire were prominent torchbearers of Enlightenment literature and philosophy.

Larry Bell

French writer Rousseau (1712-1778) was a strong advocate for reform on behalf of social empowerment and democracy which remained influential long beyond his lifetime. His 1762 book *The Social Contract* argued against the idea that monarchs are divinely empowered to legislate, asserting instead that only the people are sovereign to hold all-powerful rights.

Rousseau concluded, stating...

> *Let us then admit that force does not create right, and that we are obliged to obey only legitimate powers.*

Here, the ability to coerce is not a legitimate state power, and there is no rightful duty to submit to it.

Voltaire (1694-1778) was, in fact, a pen name of Francois-Marie Arouet. He likely used this pseudonym device to shield him from persecution for pointedly barbed criticisms against the Roman Catholic Church which he reviled as intolerant, backward and too steeped in dogma.

In keeping with many other Enlightenment thinkers of his era, Voltaire condemned injustice, clerical religious abuses, and while believing in a supreme being, regarded formalized religion as superstitious and irrational.

Voltaire vigorously emphasized empirical natural science that served in his mind as a necessary antidote to vain and fruitless philosophical investigations. Politically, he despised democracy as rule by mobs and believed that an enlightened monarchy informed by counsels of the wise was best suited to govern.

Although the political climate in the American colonies was vastly different than Europe, philosophies emerging from the Enlightenment had profound influences upon the New World as well. Among these colonist leaders, Benjamin Franklin and Thomas Paine—each in their own way—took up the rational thinking mantle.

For Paine (1737-1809) the new ideas in Europe likely prompted a desire to regard the colonies separate and independent from the British Crown. His *Common Sense*, an impassioned yet well-reasoned plea for independence, was instrumental in gathering supporters to this cause with the rallying cry of "No Taxation without Representation."

Benjamin Franklin (1706-1790) adopted a more utilitarian philosophy. While recognizing a need to become independent of the British Empire, he also foresaw the difficulties of forging a strong and lasting union out of disparate and competing colonial interests.

In 1757, Franklin was delegated to go to England as an agent of the Pennsylvania Assembly with the purported purpose of persuading the family of William Penn, as the proprietor of Pennsylvania, to allow the Colonial Legislature to tax its un-granted lands. The mission's real aim, however, was to oust the family from power, and to make the colony a royal province.

Franklin then spent the next 18 years in London, influencing fellow colonists and British sovereigns alike to suspect disloyalties to both camps. His public persona as a royalist became reinforced through privileged political British connections which enabled him to have his son, then age 31, appointed Royal Governor of New Jersey.[217]

As the eighteenth century drew to a close, passionate calls for social reform and a utopian, egalitarian society ebbed. Nevertheless, the world of Western and Colonial thought had been transformed. Science had been propelled forward by that time such that traditional authority of the Church was in real jeopardy. Monarchs no longer ruled by Divine Right, and common citizens had opened frank conversations and engagements influencing governance policies and the course of global events.[218]

If there was a historical moment that can be said to mark

the beginning of the end of the Enlightenment, it was the French Revolution. France in 1789 had devolved into anarchy where sadism perpetrated by French citizens on each other was anything but enlightened.

The French Revolution led to the rise of Napoleon a decade later.

Enlightenment ultimately gave way to 19th century Romanticism when many poets and philosophers turned away from deductive science to emphasize knowledge and imagination gained through human intuition and emotion. Prevalent themes of Romantic literature included the celebration of nature and sublime beauty, the idealization of rural lifestyles, and the rejection of rationalism, social convention, organized religion and industrialization.

British romantic poet George Gordon "Lord" Byron (1788-1824) promoted defiance, rebellion, noble deeds and contempt for tradition. Although Americans Henry David Thoreau and Ralph Waldo Emerson rejected commercialism and championed personal spiritualism, they more readily accepted science than their European counterparts.[219]

Bertrand Russell referred to Romanticism as a "cult of emotions." Yet as history continues to demonstrate, in addition to great art and literature, emotional passions also drive revolutionary scientific and technological discoveries and developments.[220]

As Wane Bundy observes in his book *Out of Chaos: Evolution from the Big Bang to Human Intellect,* a struggle between irrational and rational thinking may be an essential aspect of progressive civilization:

> *By indirection, irrationality may promote new, useful approaches to problems—finally reconciled by rationality. Ambivalence seems an innate condition of the human brain and the way*

of nature. By struggling with the extremes, our minds become informed and prompted toward the most workable solution, sometimes toward the strongest bias.

Bundy concludes:

Perhaps the most prominent example is the struggle between religion and science. [221]

An Industrial Revolution

Wayne Bundy points to modern civilization through the Scientific Revolution and the Enlightenment to the present as a revival of Ancient Greek thought. Much appreciation also is owed to powerful lessons and new ways of thinking advanced by Copernicus, Galileo, Descartes and other great minds who awakened and enabled another radical revolution of human industry, a new age of machines.

Near the 18th century's end, a rapid transition from dependence upon small craft shops using hand-production methods to the establishment of new technologies, economies and lifestyles rapidly emerged in Europe and America. Generally dated from about 1760 to sometime around the mid-1800s, the early Industrial Revolution featured mechanization of textile production.

A second phase leading to an unprecedented rise in European population and economic growth began after about 1870. This Second Industrial Revolution featured new steel making processes, large-scale manufacture of precision machine tools and the use of increasingly advanced machinery in steam-powered factories.[222, 223]

Larry Bell

Transformations of Textile Production

The Industrial Revolution began in Great Britain, then the world's leading commercial nation, a global trading empire with substantial control over North American colonies and the Caribbean. Britain also exercised significant trade influence over the Indian subcontinent through political ties with the powerful East India Company.[224]

Following the early 16th century discovery of a trade route to India around Africa, the Dutch established the East India Company and other smaller companies to engage in trade throughout the Indian Ocean region and North Atlantic Europe. Cotton textiles purchased in Eastern India and sold in Southeast Asia comprised one of the largest segments of this commerce. Cloth represented more than three-quarters of all the East India Company's exports by the mid-1760s.[225]

Sometime after 1000 AD, hand-manufactured cotton textiles had already become a major trade industry in tropical and subtropical regions in parts of India, China, Central America, South America and the Middle East. Cotton cloth could be used as a medium of exchange almost everywhere.

Europe depended upon favorable growing conditions on southern colonial plantations for cotton imports. However, the raw material was costly due to difficulties in removing seeds, putting British textile producers at a trade disadvantage with Indian cloth-goods.

In 1794, U.S.-born inventor Eli Whitney (1765-1825) developed a revolutionary machine that radically changed this trade balance. His patented cotton gin dramatically sped up the process of seed removal by applying a combination of wire screen and small wire hooks to pull cotton fibers through the device as brushes continuously removed lint to prevent jams.

A person using the new cotton gin could remove as much seed in one day as previously required two months for an individual to hand-process. The device increased the productivity of removing seed from cotton by a factor of 50. As a result, cotton had become America's leading export to supply Europe's need for raw textile material by the mid-19th century.[226, 227]

The Industrial Revolution also mechanized the spinning and weaving of cloth which had traditionally been accomplished as a cottage industry principally for domestic consumption. Home-based workers produced goods under a putting-out contract with merchant sellers who typically provided the raw cotton materials.

Farmers' wives conventionally did the spinning off-season, while the men did the weaving. Using the spinning wheel, it took between four to eight spinners to supply one handloom weaver.[228]

A flying shuttle, patented in 1733 by John Kay in England, along with later improvements, doubled the output of the weaver. His invention also worsened the imbalance between spinning and weaving.

An early spinning breakthrough occurred in 1770 with British inventor James Hargreaves' patented the spinning jenny. The device worked in a similar manner to the spinning wheel by first clamping down the fibers, then by drawing them out, followed by twisting. However, the spinning jenny produced a lightly twisted yarn only suitable for weft (the transverse thread drawn through and inserted over-and-under the longitudinal threads on a frame or loom—the warp).

A spinning frame, or water frame patented in 1769 by Richard Arkwright was able to produce a hard, medium count thread suitable for warp, finally enabling mechanically-assisted 100% cotton cloth to be made in Britain.

Samuel Compton's spinning mule introduced in 1779 yielded finer thread than hand-spinning, and at a much lower cost. This finally enabled Britain to produce highly competitive yarn in large quantities. The device combined features of the spinning jenny and water frame in which spindles were placed on a moving carriage. The system went through an operational sequence during which the rollers stopped while the carriage moved away from the drawing roller to finish drawing out the fibers as the spindles started spinning.

Entrepreneur Richard Arkwright brought ongoing advancing cloth production processes together in a mechanized cotton mill factory. Other inventors increased the efficiency of the individual steps of spinning (carding, twisting, spinning and rolling) so that the supply of yarn increased greatly.

Mechanized cotton spinning powered by steam or water increased the output of a worker by a factor of around 500. A power loom alone increased the output of a worker by a factor of over 40. And while large production gains also occurred in spinning and weaving of wool and linen, they were not nearly as great as those in cotton.[229]

Development and Benefits of Steam Power

Although the development of the stationary steam engine became very important to the Industrial Revolution, most of the power during the early period was supplied by water and wind. Steam power only underwent rapid expansion after 1800.

London-based inventor and entrepreneur Thomas Savery patented and constructed the first commercial steam power in 1698. The low-lift one-horsepower system combined a vacuum and pressure pump used in various water works and mine water-removal applications.

The first successful piston steam engine was introduced in Britain by Thomas Newcomen sometime before 1712. Its principal uses were to drain previously unworkable deep mines and to power municipal water supply pumps.

Fundamental steam engine improvements were introduced by Scotsman James Watt and business partner Englishman Matthew Boulton in 1778. Closure of the upper part of the steam cylinder redirected low-pressure steam to drive the top of pistons rather than venting it into the atmosphere as Savery and Newcomen had done. Use of a steam jacket and a separate steam condenser chamber also did away with the cooling water that had previously been injected directly into the cylinder, wasting steam in the process.

Evolutionary steam engine efficiency improvements resulted in enormous fuel savings, amounting to three-quarter or more reductions in coal use per horsepower-hour over Newcomen's.

Adaptation of stationary steam engines to rotary motion made them suitable for industrial uses. Key among these were for the development and mass production of precision machine tools, such as the engine lathe, planing, milling and shaping machines. Powering by these engines enabled all the metal parts to be easily and accurately cut, which in turn, made it possible to build larger and more powerful engines.

Improved power-to-weight ratios of new high-pressure steam engines made them suitable for mobile transportation applications. Providing lighter weight and smaller size for a given horsepower than stationary systems was accomplished by exhausting used steam directly into the atmosphere, thus doing away with a condenser and cooling water.[230]

Iron Production Advancements

Widespread railroad development after 1800 made possible by steam engine advancements was also greatly enabled by major iron production innovations needed to create the many miles of tracks along with other essential products needed to support a growing population and industrial economy. Included was bar iron used as the raw material for making hardware goods such as nails, wire, hinges, horseshoes, wagon wheel rims, chains and structural shapes.

A small amount of bar iron was converted into steel. Most of the early cast iron was refined and converted to bar iron as well, although with substantial inefficiencies.

A major improvement in iron production efficiencies during the Industrial Revolution resulted from the replacement of wood with coal, which was more abundant and less expensive. Coal required much less labor to mine than that involved in cutting wood and converting it to charcoal. In addition, other applications such as construction were causing wood to become increasingly scarce.

Another factor limiting the iron industry prior to the Industrial Revolution was a scarcity of water power to power blast bellows. In addition, the leather used in those bellows was expensive to replace.[231]

Iron master John Wilkinson patented a high-pressure hydraulic-powered blowing machine to blast air in 1757 that solved both problems. The design was later improved by making it double-acting, which allowed higher blast furnace temperatures.

The substitution of coke (conversion of coal by heating it in the absence of air) for charcoal greatly lowered the fuel cost of crude pig iron and wrought iron production. Using coke enabled economies of scale afforded by these larger blast furnaces.[232, 233]

Henry Cort developed two significant iron manufacturing advancements in, rolling (1783) and puddling (1784). The rolling mill was 15 times faster than hammering wrought iron.

The puddling process produced a structural grade iron at a lower cost than forging by means of decarburizing (or decarbonizing) molten pig iron by slow oxidation in a furnace. This remained extremely hot, backbreaking work which involved manually stirring the material with a long rod. Few puddlers reportedly lived to reach the age of 40.

Hot blast patented by James Beaumont Neilson in 1828 greatly increased fuel efficiency in iron production in the following decades. Attributed by some as one of the most important developments of the 19th century, it saved energy in making pig iron by using preheated combustion air, reducing fuel consumption by one-third using coke or by two-thirds using coal.[234]

Steel was an expensive commodity prior to the Industrial Revolution. Accordingly, it was used only where iron would not do, such as for cutting edge tools and for springs. A crucible technique developed by Benjamin Huntsman in the 1740s enabled large-scale production of cheaper iron and steel which aided a number of industries. Included were commodities such as nails, hinges, wire and other hardware items.

Perhaps most impactful, new power and metallurgical advancements enabled the development and mass production of precision machines.

Precision Machines

Pre-industrial machinery was built by craftsmen—millwrights built water and windmills, carpenters made wooden framing and smiths made metal parts. As the Industrial Revolution progressed, ever cheaper and more precise machine-made tools and metal parts became increasingly common.

The first machine tools included the screw-cutting lathe, cylinder boring machine and the milling machine. These led to capabilities enabling the economical manufacture of large numbers of precision threaded metal fasteners such as screws, bolts and nuts.

In the 1770s, Henry Maudslay built a lathe which could cut machine screws of different thread pitches. These were the first machines for mass production capable of making components with a high degree of interchangeability.

The concept of interchangeable parts first took ground in the firearms industry when French gunsmith Honoré LeBlanc promoted the idea of using standardized gun parts. Before this, individual firearms which were made individually by hand varied slightly from one to another. Thus, each weapon was unique and could not be easily fixed if broken.

It wasn't until cotton gin inventor Eli Whitney introduced the idea in the United States Department of War in the 19th century that development of interchangeable parts for small firearms really took off. Whitney had trained a large unskilled workforce using standardized equipment to produce large numbers of identically replaceable gun parts at a low cost and within a short amount of time.[235]

In the half century following the invention of the fundamental machine tools, the industry became the largest value-added industrial sector of the U.S. economy.

In 1901, Ransom Olds created and patented the assembly line, a factory process which allowed his car manufacturing company to increase output by 500 percent in one year. A Curved Dash model was able to be produced at what then was an exceptionally high rate of 20 units per day.

Henry Ford improved upon Olds' assembly line concept by using the moving platforms of a conveyor system. The vehicle chassis was towed by a rope that moved it from station to station, allowing a progressive sequence of stationed workers to assemble each part.

Ford's revolutionary assembly method enabled a Model T to be produced every ninety minutes, totaling nearly two million units in one of their best years. Often credited as the father of the assembly line, he would be more appropriately characterized as the father of automotive mass production.

Revolutionary Contributions and Consequences

Some economists, such as Robert E. Lucas, Jr., say that the real impact of the Industrial Revolution was that...

> *...for the first time in history, the living standards of the masses of ordinary people have begun to undergo sustained growth...Nothing remotely like this economic behavior is mentioned by the classical economists, even as a theoretical possibility.*[236]

The Industrial Revolution was the first period in history during which there was a simultaneous increase in both population and per capita income.

During the Industrial Revolution, life expectancy increased dramatically. The percentage of children born in London who died before the age of five decreased from 74.5% in 1730—to 31.8% in 1810-1829.[237]

Until about 1750, in part due to malnutrition, life expectancy in France was about 35 years and about 40 years in Britain. The U.S. population at the time was adequately fed, much taller on average and had a life expectancy of 45-50 years.

A very major contribution of the Industrial Revolution was to food abundance essential to nourish growing populations over the past 200 years constituting what can legitimately be termed a Second Agricultural Revolution.

As Yuval Noah Harari points out:

> *Machines such as tractors began to undertake tasks that were previously performed by muscle power, or not performed at all. Fields and animals became vastly more productive thanks to artificial fertilizers, industrial insecticides and an arsenal of hormones and medications. Refrigerators, ships and airplanes have made it possible to store produce for months, and transport it quickly and cheaply to the other side of the world. Europeans began to dine on fresh*

Larry Bell

Argentine beef and Japanese sushi.[238]

Harari and others also appropriately argue that while the growth of the economy's overall productive powers was unprecedented during the Industrial Revolution, living standards for many workers were very low. Histories of these early times bring to mind prevalent images of urban landscapes dominated by smoking chimneys and the sad plight of exploited coal miners sweating in the bowels of the Earth.

Living standards and health gradually yet dramatically improved during the 19th and 20th centuries. Labor laws, for example, addressed exploitive working conditions and compensation practices and public health acts regulated industrial sewage disposal.

And while new and more efficient machines and processes of the Industrial Revolution yielded an explosion in human productivity, for a great many—craft and farm workers in particular—it cost them their jobs and livelihoods.

The mechanization movement started first with British lace and hosiery workers, then rapidly spread to other areas of the textile industry.

In 1811, angry mobs of newly unemployed weavers and other workers turned their animosity towards attacking machinery and factories that had taken their jobs. Riots by self-identified as Luddites, supposedly followers of Ned Ludd, a mythical folklore figure, often turned violent. Many were arrested by British militia troops hired to protect industry, tried and jailed. Some were even hanged.

Unrest also occurred in other mechanized industry sectors. In the 1830s, for example, agricultural laborers in southern Britain destroyed threshing machines and burned hay bales.

Nevertheless, despite employment disruptions and workforce shifts, the Industrial Revolution created far more jobs than casualties. The abundance of more affordable products in combination with rapid increases in general consumer prosperity gave birth to a new capitalistic era of entrepreneurship, innovation and global commerce.

Liberalization of trade from an expanding merchant base allowed Britain to produce and use emerging scientific and technological developments more effectively than countries with stronger monarchies, particularly China and Russia.

Philosopher Karl Marx had predicted that capitalism would be overthrown by communism so that oppressed workers would finally be free. History didn't turn out that way.

Karl Marx got it exactly backwards.

Larry Bell

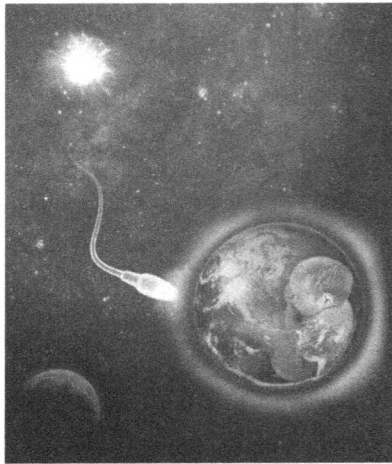

Part Nine: New Wonders and Worlds of Discovery

SCIENTIFIC AND TECHNOLOGICAL progress has advanced at an ever-accelerating pace, where inventions continuously spirit, enable and ultimately multiply new innovations and knowledge exponentially.[239]

A fast-paced 19th-20th century era of scientific discovery and technological progress began in 18th century BC Greece. From there it was rediscovered and rekindled during the 14th-17th century Renaissance, was catalyzed and objectified during the 17th-18th century Scientific Revolution, was culturally inspired during the 18th century Enlightenment, and was accelerated to warp-speed during the 18th-19th century Industrial Revolution.

The 18th-century invention of the steam engine rapidly transformed industries, railroads and later, along with the electricity-generating dynamo, electrified a whole new world of work and life-changing possibilities.

In 1864 James Maxwell developed a revolutionary theory of electromagnetic waves, and Marconi invented wireless telegraphy in the late 1890s. At about that same time, Wilhelm Conrad Roentgen discovered X-rays, French physicist Becquerel discovered radioactivity, Marie and Pierre Curie carried out their pioneering work on radioactivity using radium and Ernest Rutherford formulated an atomic structure theory which first described a nucleus encircled by electrons.

In the remarkably short span of a century, humans learned to harness lightning, to develop wings and to split atoms. Such discoveries and advancements have at once and forever transformed society in two fundamental ways. Just as they continue to represent exciting forces of promise, those same forces empower terrifying weapons of war.

Electrifying Society

Electrification enabled by the invention of electromagnetic generators—powered by innovations of industrial-scale steam turbines—must certainly be credited as one of the top-transformative 20th-century developments. The original operating principle, now known as Faraday's Law, was discovered by Michael Faraday in the years 1831-1832, namely that an electromagnetic force is generated in an electrical conductor which encircles a varying magnetic flux. Faraday's first electromagnetic generator used a copper disc rotating between poles of a horseshoe magnet to produce a small DC voltage.

The first electric generator capable of delivering commercially practical power was the dynamo used in 1844 for electroplating. Modern dynamos capable of producing industrial-scale electricity were invented independently two decades later by Sir Charles Wheatstone, Werner von Siemens and Samuel Alfred Varley between 1866 and 1867. Siemens' design, which incorporated electromagnets rather than permanent magnets, greatly increased the dynamo output required for high power-demand applications such as electric arc furnaces used in the production of metals.[240]

The first power stations supplied Direct Current (DC) which was well-suited for a number of applications such as electric street railways, machine tools and certain industrial applications where speed control was important. Serbian-American electrical engineer Nikola Tesla's invention of Alternating Current (AC) soon became the option of choice for general electrification because it could be transformed to high voltages with low power losses and also enabled motors to run at very constant speeds.

Although Tesla had attended the Austrian Polytechnic in Graz, Syria, on a scholarship, he left after his second year in 1881 to work as a low-wage draftsman at the Central Telegraph Office in Budapest, Hungary, where he was soon promoted to a chief electrician position. Two years later Tesla moved to America and was employed by Edison's Machine Works to develop a high-voltage arc lamp lighting system.

Recognizing that Edison's DC technology was incompatible with high-voltage requirements, Tesla proposed a revolutionary AC alternative which was rejected. He then left Edison's company in 1885 after only six months, and together with some investors, founded the Tesla Electric Light & Manufacturing Company. After proceeding to patent a new arc lighting concept along with new types of AC motors and electrical transmission equipment, the enterprise folded. Tesla lost control of the patents, leaving him broke.

In 1887, together with two new investors, Tesla then formed the Tesla Electric Company. The new company developed an AC induction motor, a concept affording large advantages for long-distance, high-voltage transmission. Engineers at Westinghouse Electric & Manufacturing recognized the importance of the design, and the company negotiated a licensing deal. A Westinghouse-General Electric merger arrangement later purchased the patent from Tesla's company.

Tesla's achievements following that period were transformative. He accomplished the first successful wireless energy transfer to power electronic devices in 1891, conducted the earliest demonstration of fluorescent lighting, and influenced the development of modern electrical generators and turbine designs.

In 1893, the Westinghouse Electric Company implemented Tesla's AC system to light the World Columbian Exposition in Chicago. The demonstration proved to be more efficient than the direct current system marketed by Edison, and rapidly became the basis for most modern electric power distribution systems. In 1895, Tesla and Westinghouse developed the world's first hydroelectric power plant at Niagara Falls.

At the turn of the century, Tesla set up a laboratory in Shoreham, Long Island featuring a Wardenclyffe Tower project intended to provide intercontinental wireless communications as a more powerful transmitter in competition with a Marconi radio-based system which Tesla regarded as a copy of his design.

Tesla's investors dropped out after Marconi's system won out in December 1901 by successfully transmitting the letter "S" from England to Newfoundland. He died virtually penniless following an unsuccessful attempt to sue Marconi for infringement on his wireless patents.

Thomas Alva Edison, Tesla's earlier rival in the AC-DC electric current war, is recognized as one of America's most prolific inventors. His more than 1,000 patents include such innovations as incandescent electric lights, the microphone, telephone receiver, stock ticker, phonograph, movies and office copiers.

At age 20, Edison secured work in Cincinnati, Louisville, Indianapolis, Memphis and Boston as an itinerant Western Union telegraph operator. The job suited his interest in learning more and more about telegraphy, including how to improve the equipment.[241]

By 1969, Edison's entrepreneurship as an inventor began to really take off. His patent applications included a telegraphic stock ticker which became standard office equipment in America and Europe, and a printing telegraph for gold bullion and foreign exchange dealers. He also figured out how a central telegraph office could control the performance of equipment from remote locations and developed a method to transmit as many as four messages over a single wire.[242]

On July 18, 1877, as Edison tested an automatic telegraph which had a stylus to read coded indentations on strips of paper, the friction revealed an unexpected hum that attracted his attention. As Douglas Tarr at the Edison National Historical Site in West Orange, New Jersey reported:

> *Edison seemed to reason that if a stylus going through indentations could produce a sound unintentionally, then it could produce a sound intentionally, in which case he should be able to reproduce the human voice...A talking machine!*[243]

Edison worked on and off over more than two decades to advance that concept to do much more than just talk. His innovation ultimately produced sound quality that brought high fidelity music to homes of world audiences.

In 1879, Edison's Menlo Park laboratory demonstrated the first high-resistance incandescent light which passed electricity through a thin platinum filament in a glass vacuum bulb to delay melting. After the original model worked only for an hour or two, Edison went on to try carbonized filaments made of almost every imaginable plant material, including some specially ordered fibers from the tropics. The best performer proved to be carbonized filaments of common cotton.[244]

Many more innovations followed during the late 1880s and

early 1890s. In the area of photographic optics, for example, Edison demonstrated the potential of using tough, flexible celluloid motion picture film, worked out mechanical problems of advancing the film steadily across a photographic projection lens without tearing and linked a new motion picture camera with an improved phonograph featuring synchronized sound, producing the Kinetoscope that projected talking images on screens.

Edison's legacy of achievement is commemorated by numerous companies that bear his name. Included are: Edison General Electric (which merged with the Thomson-Houston electric company to form General Electric); Commonwealth Edison (now part of Exelon); Consolidated Edison; Edison International; Detroit Edison (a unit of DTE Energy); the Edison Electric Institute (a trade association); the Edison Ore-mining Company; the Edison Portland Cement Company; Ohio Edison (which merged with Centerior in 1997 to form First Energy); and Southern California Edison.

Splitting Light and Atoms

It's difficult to imagine anyone who exemplifies a greater genius in the popular minds of most people than Albert Einstein. The products of his thinking delivered far more than he originally advertised in a 1905 letter to his friend Conrad Habicht.

Einstein, then working as a low-level patent examiner, wrote:

> *I promise you four papers. The first deals with radiation and the energy properties of light and is very revolutionary, as you will see if you send me your work first.*

That paper postulated that light could be regarded both as a wave as well as a stream of tiny particle packages called quanta.

Einstein went on to say:

> *The second paper is a determination of the true*
> *sizes of atoms... The third proves that bodies on*
> *the order of magnitude 1/1000 mm, suspended*
> *in liquids, must already perform an observable*
> *random motion that is produced by thermal*
> *motion. Such movement of suspended bodies*
> *has actually been observed by physiologists who*
> *call it Brownian motion.*

Using statistical analysis of random collisions, that third paper established the true existence of atoms and molecules.

Einstein continued that:

> *The fourth paper is only a rough draft at this*
> *point, and is an electrodynamics of moving*
> *bodies which employs a modification of the*
> *theory of space and time.*

This later became famously known as the Special Theory of Relativity.

That same year, he was also working on a short addendum to that fourth paper which drew a relationship between energy and mass. The addition envisioned bending of light beams and warping of space. That relationship is briefly and most famously of all summarized as $E=mc^2$. His predictions of how much gravity actually bends light were later validated during a 1919 solar eclipse.

In 1895, 16-year-old Einstein imagined what it would be like to ride alongside a light beam. A decade later, this boyhood musing provided the conceptual foundation for two great advances of 20th-century physics: relativity and quantum theory.

Then in 1915, only one more decade after that, he followed that light beam of imagination to produce his everlasting crowning scientific accomplishment. That General Theory of Relativity explained how space-time is warped by an interplay between matter, motion and energy.

Einstein likened this circumstance to rolling a bowling ball onto the two-dimensional surface of a trampoline. Then when some billiard balls are added, they move toward the bowling ball not because it exerts some mysterious attraction, but rather, because of the way it curves the trampoline fabric.

Here, space and time are not two separate things, but together form space-time where energy and mass are actually different forms of the same thing. How these mass versus energy determinations are measured is influenced by how fast the object and observer are moving relative to one another.[245]

A February 2016 announcement which the Royal Swedish Academy accurately described as "a discovery that shook the world" affirmed that Einstein had been proven right. Just as his 1916 General Theory of Relativity had predicted, sensitive Earth-based instruments recorded that gravity waves emanating from the collision of two black holes a billion light years away jiggled space-time with invisible cataclysms which reached us.

That faint "chirp" signal which was received at separate facilities in different states lasting only a fifth of a second was greeted by thousands of scientists as a loud opening bell for a whole new era of astronomical revelations.

Instruments at the U.S. Laser Interferometer Gravitational Observatory (LIGO) detected ripples in the space-time grid produced by a different type of event on August 17, 2017, which recorded the collision of two neutron stars. In addition to gravity waves, the spectacle released visible light which was observed by Earth-based telescopes. Initially appearing as a bright explosion of blue, the color soon faded to a deep red.

The discovery confirmed, as expected, that collisions of

neutron stars produce enormous gamma-ray bursts, along with about half of all heavy elements which are dispersed in gases that eventually settle down and condense to form new stars and planets.

Einstein was intellectually absorbed with a "wave-particle-paradox" whereby light can be measured either as waves of light or as energy particles depending upon which equipment we select to observe it.

It should be noted that those waves of light can also be measured as energy particles which don't contain any physical stuff. Depending upon which equipment we select to observe it, some experiments show that light is wave-like, while others show that it is a particle-like phenomenon.

Thomas Young's 1903 experiments showed that light must be wave-like, while Einstein proved that it is particle-like.

Einstein's theory proposed that light is comprised of tiny particles (photons) analogous to a stream of bullets, whereby energy itself, is quantized. He termed this a photoelectric effect.

Max Planck, the first physicist to calculate the sizes of energy packets (quanta) in various waves of light frequency (color) using his mathematical invention famously known as Planck's constant. All of those packets of color, red for example, have the same size.

As Planck described Einstein's theory:

> ...the photons (the 'drops' of energy) do not grow smaller as the energy of the ray grows less; what happens is that their magnitude remains unchanged and they follow each other at greater intervals.[246]

Einstein was not able to dispute the contradiction between light as a wave versus light as quanta, but simply took the contradiction as something which would probably be understood later. Nevertheless, while he is far more famous for two revolutionary theories of relativity, both were based upon his discoveries regarding the quantum nature of light which earned him a Nobel Prize.

Although Einstein never embraced what came to be recognized as a scientifically well-established yet counterintuitive quantum mechanics theory, he is credited with advancing, neither could he dispute that it invariably worked. For example, as discussed later in this book, principles of quantum theory are now being applied to create advanced computers with astounding processing capacities.

Niels Bohr, a Danish physicist who earned a Nobel Prize in 1922 for his contributions to quantum mechanics, argued famously with Einstein on the subject. Einstein lamented, "Alas, our theory is too poor for experience." Whereas Bohr replied, "No, no! Experience is too rich for our theory." [247]

While Newtonian physics works wonderfully well to describe and predict events in our everyday world, it cannot account for phenomena in the subatomic realm which appear to be governed by very different rules.

Just as Einstein's breakthrough Special Theory of Relativity affirms, appearances of observed subatomic events (such as light effects) are relative and dependent upon the observers. Since atoms are far too small to actually see, all that scientists can do is speculate about what is there based upon certain observations regarding how atoms appear to behave.

Quantum mechanics takes this condition one very bizarre step farther. The very fact of being observed, and by whom, influences the very event being witnessed.

The new quantum theory model presents a vision of a subatomic world comprised of unimaginably small particles which have no material substance, yet for convenience, are statistically measured as quanta in terms of energy units in the same way as particles. These quanta unceasingly change measurable appearances from energy to mass and back, although "within a common identity."

Considering size distance comparisons between atoms and subatomic particles versus between our Solar System and planets, for example, distances between an atomic nucleus and its electrons are far greater.[248]

As described by science writer Gary Zukav, the difference between the atomic level and subatomic level is as great as the difference between the atomic level and the entire planet.[249]

For another comparison, Zukav asks us to imagine an atom as the size of a grain of sand in the center of the dome of Saint Peter's Basilica in the Vatican, with electrons the size of dust particles revolving around its outer edge. However, unlike dust particles which can be visualized as things, quantum mechanics views subatomic particles only as tendencies to exist or tendencies to happen which can only be seen in the form of mathematical probabilities.

In addition to revolutionary contributions to sciences at all scales—ranging from the Universe to quantum subatomic quanta—Einstein's work also led to many important technological advancements by others. Included are photoelectric cells, lasers, fiber optics, semiconductors and nuclear power.

Regarding the latter, his discovery that $E=mc^2$ (where energy is proportional to mass multiplied by the extraordinarily huge number of the speed of light squared) is one of the most consequential scientific game-changers in human history. While that equation appears to be remarkably short and simple, it has since enabled humanity to harness the power contained in tiny atoms both to power prosperity and to annihilate itself.

Humanity Takes Flight

Dreams of flight likely date back to humankind's earliest conscious fascination with the soaring freedom of birds. During the mid-1400s, Leonardo da Vinci studied the structures and workings of their wings in attempts to produce machines that might bring such fantasies to fruition through a variety of mechanical devices. One of these—a hypothetical, un-tested flapping-wing concept called an ornithopter—was unsuccessfully attempted by many other inventors over the next four centuries.

The 18th-century discovery of hydrogen gas led to the invention of tethered and free-flying balloons which were first used for military surveillance purposes, these, in turn, led to the passenger-carrying rigid dirigible balloons pioneered by Ferdinand von Zeppelin in Germany, also referred to as airships which dominated long-distance flight until the 1930s.

The catastrophic ignition of hydrogen tanks used on the German Luftschiff Zeppelin company's longest-class dirigible, the LZ 129 Hindenburg, marked the beginning of the end of the airship's popularity. The May 6, 1937, disaster which killed 36 people at the end of its first North American transatlantic flight at the Lakehurst Naval Air Station in New Jersey had been preceded by crashes of several others, three of which cost even greater numbers of fatalities.

Although non-flammable helium was known to be the safest gas for airships, it was rare, and therefore far more expensive than hydrogen. The U.S. Government issued a Helium Control Act of 1927 to ban its export, virtually forcing the use of hydrogen for large-scale lighter-than-air passenger craft.[250]

Late 19th-century experiments with heavier-than-air craft and early-20th century experiments in engine and aerodynamic technology innovations provided revolutionary foundations for modern aviation.

In 1891, American astronomer Samuel Pierpont Langley published a paper titled *Experiments in Aerodynamics* and on May 6, 1896, launched the first two sustained-flight demonstrations of an unpiloted heavier-than-air craft which he launched by a spring-actuated catapult mounted on top of a houseboat on the Potomac River near Quantico, Virginia. The longest flight of these two on that day traveled 3,300 feet at about 25 miles-per-hour at top speed.

Langley launched another successful unpiloted demonstration witnessed by Alexander Graham Bell on November 28, 1896, which traveled nearly one mile. This was followed by a quarter-scale passenger engine-powered concept version he tested in 1901 and 1903.

Sadly, for Langley, his efforts to create the first engine-powered passenger-carrying aircraft ended nine days after a second abortive attempt on December 8, 1903. The Wright brothers accomplished this feat on December 17, 1903.

Orville and Wilbur Wright had built and tested a series of kite and glider designs prior to attempting to build a powered design. After the first partial-scale glider they designed flew poorly, they built a makeshift wind tunnel to test 200 wing designs to develop a superior full-size version.

The Wrights invented an innovative wing warping concept along with a steerable rear rudder for controlled flight, along with a low-powered internal combustion engine and specially shaped wooden propellers for optimum power efficiency.

Orville's historic 12-second Flyer I flight, which took place four miles south of Kitty Hawk, North Carolina, traveled a total of 120 feet. This was followed by one flown 852 feet by his brother Wilbur that same day which lasted nearly a minute.

The brothers continued to improve and test their designs at Huffman Prairie near Dayton, Ohio. Their third version became the first practical aircraft to fly consistently under full pilot control from its starting point safely and without damage. Wilber successfully piloted a Flyer III a record-breaking 24 miles in 39 minutes, 29 seconds on October 5, 1905.

Airplanes soon gained interest for military purposes. They were first used by Italy for reconnaissance, bombing and artillery correction flights in Libya during their 1911-1912 war with

Turkey. Bulgaria followed with bombing attacks on Ottoman positions during the First Balkan War of 1912-1913.

World War I witnessed major offensive, defensive and reconnaissance airplane uses both by Allies and Central Powers. Opposing pilots began shooting at one another, and in late 1914, Roland Garros of France came up with the deadly idea of attaching a fixed machine gun to the front of his plane. The first aerial victory was scored on July 1, 1915, by German pilot Lieutenant Kurt Wintgens flying a purpose-built fighter plane featuring a synchronized machine gun.

Air-to-air combat became the making of legendary heroics. German ace Manfred von Richthofen, better known as the Red Baron, shot down 80 planes. René Paul Fonck on the Allied side was credited with 75 aerial victories.

Aircraft technology between World War I (1919) and World War II (1939) rapidly evolved from low-powered wood and fabric biplanes to sleek high-powered aluminum single-winged craft.

World War II not only rapidly increased the pace of aircraft development and production, but also that of more precise and lethal flight-based weaponry used in strategic large-scale bombing campaigns and dive bomber attacks on small targets. New technologies such as radar and communication systems for coordinated air defenses accompanied these accelerating developments.

In 1942, Germany introduced the first operational jet aircraft (Heinkel HE 178), and in 1943, also produced the first jet bomber (Arado Ar 234). Germany also developed the first cruise missile (V-1), the first ballistic missile (V-2) and the first operational rocket-powered combat aircraft (Me 163). However, late introduction, fuel shortages and a declining war industry limited overall German jet and rocket-powered aircraft advantages.

The immediate post-World War II era saw great

advancements in jet and rocket-powered flight. American Chuck Yeager broke the sound barrier in 1947 in the rocket-powered Bell X-1. Jet aircraft broke distance barriers in 1948 and 1952, first crossing the Atlantic, and then flying non-stop to Australia.

The Korean War saw extensive air-to-air combat and bombing missions. U.S. fighters are estimated to have shot down as many as 700 Soviet air-combatants. Most of these dogfights took place over enemy-controlled areas.

The Vietnam War witnessed a strategic emphasis upon combat with air-to-air missiles. Close-proximity dogfights became less frequent.

The invention of nuclear bombs increased the strategic importance of military aircraft during the Cold War between the East and West. At first, supersonic interceptor aircraft were produced in great numbers by both sides to counteract devastating threats posed by even a small fleet of long-range bombers. By 1955, this emphasis shifted to surface-to-air missiles; then later again to prioritize intercontinental ballistic missiles capable of deploying nuclear warheads.

Any discussion of important human milestone aviation innovations must also include helicopter developments. Although the original general concept dates back to Leonardo da Vinci, reliable helicopters capable of stable hover flight were developed decades after fixed-wing aircraft. This circumstance is largely due to a requirement for more power versus weight requirements. Improvements in engines and fuels during the first half of the 20th century were a critical factor in making them practical for modern warfare and civilian applications.

In 1885, Thomas Edison had attempted to build a helicopter powered by an internal combustion fueled by guncotton, an explosive. Explosions of the demonstration damaged the craft and badly burned one of his workers. Edison later patented a helicopter concept powered by a gasoline engine which never flew.[251]

Frenchman Etienne Oehmichen set an early helicopter record in 1924 with a four-rotor craft which flew 1,180 feet. German engineer Heinrich Focke designed and built the first practical twin-rotor concept which in 1937 broke all previous helicopter records.[252]

Nazi Germany developed and used small numbers of helicopters during World War II for observation, transport and medical evacuation. Extensive bombing by Allied forces limited their production capacity.

In the United States, Russian-born engineer Igor Sikorsky developed the first practical lifting helicopter design. Produced for the military primarily for search and rescue during World War II, the craft had a single main rotor, along with a smaller rotor mounted on the tail boom to counteract torque produced by the larger one.

A key helicopter technology breakthrough occurred in 1951 when Charles Kaman applied a new kind of turboshaft piston engine developed in Germany to reduce weight and improve efficient performance. The lightweight turboshaft design led to the development of larger, faster and higher-performance helicopters, while many smaller and less expensive helicopters still use piston engines.[253]

Medivac for emergency medical airlift use was pioneered during the Korean War which dramatically reduced the previous average time needed to reach a medical facility during World War II and the Vietnam War. Military applications now also make extensive uses of helicopters mounted with missile launchers and mini-guns to conduct aerial attacks on ground targets, as well as to ferry troops and supplies where the lack of an airstrip makes transport via fixed-wing aircraft impossible.

Larry Bell

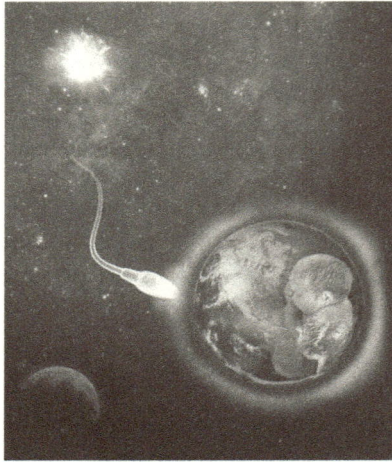

Part Ten: The World at War

TWENTIETH-CENTURY ELECTRIFICATION powered industrial mass production of new military tanks, submarines, airplanes and other armaments of World Wars I & II. The harnessing and unleashing of enormous energy stored in atoms, in combination with aerial bombers, ushered in weapons of previously unimaginable horrors of death and destruction that played a major role in ending World War II.

The development of rocketry, surveillance radar and electronic guidance systems established a new era of push-button surface-to-surface, surface-to-air and air-to-air warfare that dominated proxy wars in Korea and Vietnam.

Evolutionary rocketry evolution leading intercontinental ballistic missiles capable of delivering nuclear and thermonuclear devices to all points on the planet brought the East-West Cold War to the very brink of Mutually Assured Destruction (MAD).

Ironically, these same innovations and technologies of horrific havoc also continue to enrich and empower humanity with more abundant necessities and increasing conveniences,

232

such as electrification, food and industrial production, air travel and global Internet...to name but a few.

Supreme Commander of Allied forces during World War II, General Dwight D. Eisenhower, recognized untenable social and economic war burdens on all of humanity:

> *Every gun that is made, every warship launched, every rocket fired signifies in the final sense, a theft from those who hunger and are not fed, those who are cold and are not clothed. This world in arms is not spending money alone. It is spending the sweat of its laborers, the genius of its scientists, the hopes of its children. This is not a way of life at all in any true sense. Under the clouds of war, it is humanity hanging on a cross of iron.*[254]

World War I (1914-1918)

The First World War, also commonly known as The Great War, contributed to the fall of four venerable imperial dynasties— Germany, Austria-Hungary, Russia and Turkey. The conflict which pitted powerful forces against one another ultimately redrew much of the map of Europe.

The Central Powers were comprised of Germany, Austria-Hungary, Bulgaria and the Ottoman Empire. Allied Powers, initially known as The Triple Entente, originally comprised the British Empire, France and Russia; and later included Italy, Romania, Japan and the United States.[255]

Broadly premised upon a theory that future wars could be prevented if all ethnic groups had their own homeland, this "war to end all wars" did nothing of the kind. By the time World War I was over and Allied Powers claimed victory, more than 9 million soldiers perished, 21 million more were wounded and direct and indirect civilian casualties numbered in several additional millions of casualties.

Thanks to new military technologies and the horrors of trench warfare, World War I saw unprecedented levels of carnage and destruction. Also referred to the first modern war, it introduced the early development and mass deployment of numerous new types of weapons that continue to be in use today. Included are the machine gun, U-Boats and deadly gases by the Germans; the tank by the British; and aerial combat aircraft and armament developments by both sides.

During the early 1900s, Europe was a tinderbox of tension and military rivalry. Although there had been a number of long-existing alliances involving the European powers, the Ottoman Empire, Russia and other parties, political contentions regarding the Balkans (Bosnia, Serbia and Herzegovina in particular) threatened to disrupt fragile stability.

A flashpoint of conflict was ignited on June 28, 1914, by the assassination in Sarajevo of the Austro-Hungarian Empire's heir to the throne, Erzherzog Franz Ferdinand, by Gavrilo Princip of the Bosnian Serbs' liberation movement. Interwoven alliances and old conflicting aggressions dragged Russia and Europe into the war in support of that small Serbian state.[256]

At the start of the war, Germany stood behind its ally Austria-Hungary in a confrontation with Serbia which was under Russian protection. Russia was also allied with France.

Scarcely more than a month after Ferdinand's assassination, Germany wasted no time declaring war on Russia. Then, rather than attacking Russia, Germany instead sent its main armies through Belgium to attack Paris from the north. This invasion, in turn, caused Britain to declare war on Germany on August 4th of that year.

Other parties joined on both sides of the conflict. Turkey entered the war on Germany's side. Italy, which had previously been allied with Germany and Austria-Hungary prior to World

War I later switched to the Allied side in May 1915. The United States joined with the Allies in 1917.

Germany's primary reasons for provoking World I are varied and debated. One argument holds that Germany seized upon the Serbian conflict to advance a long-awaited opportunity to politically and economically dominate Europe. Germany was emerging as a superpower through expansion in Africa and wanted to compete on the world stage.

Some historians argue that the war was inadvertently caused by the breakdown of Alsace-Lorraine, a complex series of agreements that got out of control. Earlier events such as an Agadir crisis had put Germany particularly at odds with France over German annexation in 1871.

Still others postulate that Germany launched the offensive as a defensive preemptive strike against increasingly powerful surrounding enemies—Russia, France and Britain—who threatened to crush it. Many German nationalists saw war as inevitable and wanted to seize upon a chance to display military glory.

All of these arguments are most likely true.[257, 258]

World War I commenced on August 4, 1914, as Germans assaulted and rapidly captured the heavily-fortified Belgian city of Liege. Using enormous cannons, the most powerful weapons in their arsenal, the siege left great death and destruction in its wake. Before then advancing on towards France, German troops brutally executed many civilians, including a priest, who were accused of inciting resistance.

Germany launched an aggressive two-action-front military strategy known as the Schlieffen Plan (named for its mastermind, German Field Marshal Alfred von Schlieffen) which invaded France through Belgium in the West, while simultaneously confronting Russia in the East.

Russian forces retaliated by invading German-held regions of East Prussia and Poland, but were stopped short of victory by

German and Austrian forces at the bloody Battle of Tannenberg in late August 1914. The crushing defeat which occurred barely a month into the conflict became emblematic of the Russian Empire's World War I experience. The battle almost completely destroyed the Russian Second Army, leading to the suicide of its commanding general, Alexander Samsonov.

Russia's army continued to mount several unsuccessful Eastern Front offensives against German lines between 1914 and 1916. Successive battlefield defeats, combined with economic instability and homeland food scarcity, provoked simmering hostilities which exploded in the Russian Revolution of 1917.

In early December 1917, Russia entered into an armistice which ended all military actions against the Central Powers. Russia agreed in the armistice to recognize Ukraine independence.

Mounting war discontentment among the Russian population, especially poverty-stricken workers and peasants, became directed against the imperial regime of Czar Nicholas II and his unpopular German-born wife, Alexandra, who were all assassinated.

Spearheaded by Vladimir Lenin, the Bolsheviks negotiated a Treaty of Brest-Litovsk with Germany at a huge cost to Russia which ceded the Baltic States to Germany and its province of Kars Oblast in the south Caucasus to the Ottoman Empire.

The overthrow of the Tsarist regime and execution of His Imperial Majesty Nicholas II and his family sparked a wave of communist revolutions across Europe. However, the European revolutions were defeated, Vladimir Lenin died in 1924 and within a few years, Joseph Stalin displaced Leon Trotsky as the de facto leader of the Soviet Union.

Firmly in power, Stalin applied unchallenged autocratic rule over the Russian population and economy through widespread purges against all suspected dissenters.

Russia's withdrawal from World War I freed German

troops to face remaining Allies on the Western Front. The first of these battles had been waged between armies of Italy and Austria-Hungary on the Italian Front at Isonzo in northwest Slovenia between June 23 and July 7, 1915. German reinforcements had helped Austria-Hungary win a decisive victory.

Later, with increased British, French and ultimately American assistance, Allies had begun to take back the Italian Front.

Two of the longest and most costly battles of the Western Front campaign had been fought at Verdun in northeastern France (February-December 1916) and the Somme Offensive on the upper reaches of the River Somme in northern France (July-November 1916).

German and French troops suffered close to a million killed and wounded trench warfare casualties in each of these battles, making them among the bloodiest in human history. The Somme Offensive was the first-ever battle fought with tanks, which were then in a very early stage of development with 4 miles per hour top speed.

Prior to 1917, the United States had adopted a policy of World War I neutrality favored by President Woodrow Wilson which continued to engage in commerce and shipping with European countries on both sides of the conflict. This position became increasingly less tenable in the face of unchecked German submarine aggression against neutral ships, including those carrying passengers.

The situation came to a head in 1915, when, after declaring waters surrounding the British Isles a war zone, German U-boat submarines sunk several commercial and passenger vessels, including some U.S. ships. The widespread protest erupted over the torpedo loss of the British Lusitania ocean liner traveling from New York to Liverpool with hundreds of American passengers aboard.

In February 1917, the U.S. Congress passed a $250 million arms appropriations bill to prepare the United States for war. America declared a declaration of war after Germany then sunk four more U.S. merchant ships during the following month.

The Imperial German Navy made rapid progress closing the gap with Britain Royal Navy superiority which had been unchallenged by any other nation's fleet prior to World War I. A lethal fleet of U-boats contributed to Germany's strength on the high seas.

Nevertheless, the battle of Jutland (May 1916)—the biggest naval engagement of World War I—left British naval superiority on the North Sea intact. As a result, Germany would make no further attempts to break an Allied naval blockade over the remainder of the war.

German troops launched what would become the last German offensive of the war in the Second Battle of Marne on July 15, 1918. French troops joined by 85,000 American soldiers and British Expeditionary Forces pushed back the German offensive and successfully launched their own counteroffensive just three days later.

The Second Battle of the Marne turned the tide of war decisively towards the Allies, who then rapidly regained French and Belgium territories. Massive casualties forced Germany to call off a planned offensive in the Flanders region stretching between France and Belgium, their last hope of victory.

Facing dwindling resources on the battlefield, a distraught populace on the home front and the surrender of its allies, Germany was finally forced to seek an armistice which ended the war on November 11, 1918.

German World War I veteran Erich Maria Remarque characterizes some horrific human consequences from his individual perspective in his 1929 novel, *All Quiet on the Western Front*.

A man cannot realize that above such shattered bodies there are still human faces in which life goes its daily round. And this is only one hospital, a single station; there are hundreds of thousands in Germany, hundreds of thousands in France, hundreds of thousands in Russia. How senseless is everything that can ever be written, done, or thought, when such things are possible? It must be all lies and of no account when the culture of a thousand years could not prevent this stream of blood being poured out, these torture chambers in their hundreds of thousands. A hospital alone shows what war is.[259]

Following the defeat, while Germany was never occupied by Allied troops, the Treaty of Versailles forced them to make large payments to repair war damages and also to accept a liberal democratic government imposed on it by the victors following forced abdication of Kaiser Wilhelm.

Saddled with heavy reparations and denied entrance into a new League of Nations which was formed to mediate disputes and prevent future wars, Germany felt tricked into signing the treaty. Long-smoldering German hatred for the Versailles terms and its authors can be counted among causes which led to World War II.

World War I had many other important historical consequences.

Despite the Turkish victory at Gallipoli, later defeats by invading forces and an Arab revolt had combined to destroy the Ottoman economy and devastate its land. The Turks signed a treaty with the Allies in late October 1918.

Dissolving from within due to growing nationalist movements among its diverse population, Austria-Hungary reached an armistice on November 4. New states including

Yugoslavia and Czechoslovakia were created from the former Austro-Hungarian Empire to accommodate nationalist aspirations.[260]

World War I also brought about other global treaties and policy agreements. For example, terrifyingly severe effects that chemical weapons such as mustard gas and phosgene had on soldiers and civilians galvanized public and military attitudes against their continued use. Geneva Convention agreements banning the use of chemical and biological agents in warfare remain in effect today.

Unfortunately, as for being The War to End All Wars, this was not to be the case. A century of devastating 20th-century conflicts had only begun.

World War II in the European Theaters

Adolf Hitler came to power in 1933 dedicated to restoring Germany's devastating losses of honor, prestige and territories resulting from World War I defeats.

He brought with him a variant of fascism called Nazism, which was directed to the goals of annexing Central and Eastern Europe as vassal states and subjecting Slavic populations to serve slave labor and economic interests of a German Herrenvolk, or a racially-pure master race. Jews were targeted for genocide as Untermensch (subhuman).[261, 262]

At its core, Hitler's World War II holocaust was an atrocity against humanity in which Nazi Germany systematically murdered some six million people—approximately two-thirds of the Jewish population in Europe. Other groups were also targeted for slaughter, including Slavs (chiefly ethnic Poles, Soviet war prisoners and citizens); those deemed incurably sick; political and religious dissenters (including communists and Jehovah's Witnesses); and homosexual men. Altogether, as many as 17 million deaths resulted.

Hitler's Nazi government began to exclude Jews from German civil society soon after taking control. Boycotts of Jewish businesses were organized, and Nuremberg Laws passed in 1935 excluded German Jews from Reich citizenship and prohibited them from either marrying or having sexual relationships with persons of "German or related blood."

In 1933, Nazis had already begun building a network of concentration camps to imprison political opponents and others charged as undesirables. By 1939 the regime had established more than 42,000 camps, ghettos and other detention sites across Europe.

Jewish deportation to the ghettos culminated in the policy of extermination the Nazis called the Final Solution to the Jewish Question which was discussed by senior Nazi officials in January 1942 at the Wannsee Conference in Berlin. Resultant mass killings by gunshot and poison gas continued until the end of World War II in Europe in May 1945.

Nazism was not the only form of savage fascism at that time. Nearly all new democracies in Eastern European nations had collapsed and been replaced by authoritarian regimes. Benito Mussolini ruled Italy as Prime Minister from 1922, then dropped the pretense of democracy to become a paramilitary fascist (and fiercely anti-communist) dictator in 1925. Spain became a repressive dictatorship under General Francisco Franco in 1939 after the Spanish Civil War until the time of his death in 1975. Stalin rose to communist dictator status in 1929 to enact a Great Purge of political dissidents during the 1930s.

Inflamed by the aftermath of World War I humiliations, Hitler harbored deep animosity against communism in general, and a Russian menace in particular. Seeing no national threat to themselves, many leaders in Western Europe and the United States greeted his rise to power with broad ambivalence. Desirous of no new war engagements, America was preoccupied with combating a widespread Great Depression (1929-1939).

Larry Bell

Fascism first appeared in Italy with the rise of Benito Mussolini in 1922. The ideology was supported by a large proportion of the upper classes as a strong challenge to the threat of communism.

When Adolf Hitler came to power in 1933, a new variant of fascism called Nazism took over Germany and ended the German experiment in democracy.

Hitler initiated an ambitious global domination plan through the German annexation of Austria in 1938. This was followed by the annexation of German-speaking Sudetenland in western Czechoslovakia which he negotiated with France, Great Britain and Italy through the threat of war at the Munich Conference.

In what has since come to be known as the Munich Betrayal, Hitler pledged what would be his last territorial claim in Europe. Eager to avoid military conflict with Germany, Britain and France determined that it was in their best interests to trust Hitler's assurance that he would protect the security of the Czech state.

Sudetenland was of special strategic importance to Czechoslovakia because most of its border defenses against Germany were located there. Nevertheless, Czechoslovakia was not invited to attend the emergency meeting.

A low-intensity, undeclared German-Czechoslovak war already occurring at the time. Meanwhile, Poland was also relying on a separate nonaggression pact with Germany to annex bordering Czechoslovak territories for itself. Facing combined German and Polish threats along most of its border, with a major part of the remaining border being with shared with Hungary, Czechoslovakia yielded to French and British diplomatic pressure to establish a new border with Hungary. Consequently, control of Czechoslovakia was ceded to Germany.[263, 264]

Soon after issuing deceptive non-aggression promises to Czechoslovakia, Britain and France, Hitler also issued empty assurances of protection to Poland. Then, on September 1, 1939,

Nazi Germany unleashed a lightning war Blitzkrieg against all of them which commenced World War II. Acting in a secret alliance with Hitler, Stalin enacted a Soviet Union attack on Poland from the East only sixteen days later. Polish forces were soon overwhelmed, and its government leaders fled to London.[265]

The Germany-USSR Molotov-Ribbentrop Pact gave Stalin free rein to take control of Eastern Poland as well as the Baltic republics of Estonia, Latvia and Lithuania. According to the secret treaty, all of these territories were to remain in Soviet possession after the war. Stalin also launched an attack on Finland, which met stiff resistance and gained only limited ground. The action later backfired, prompting the Finns to ally with Germany to attack the Soviet Union in 1941.[266]

Germany began World War II with a new type of warfare that was very different than the trench fighting of WWI. Although much-upgraded tank groups called panzer divisions were used in support of highly-mobile infantry, the strategy also employed massive and relentless air attacks.

On May 10, 1940, Germany launched a massive Western Front assault on Belgium, the Netherlands and Luxembourg. Belgium's King Leopold surrendered that same month, exposing the entire flank of Western Allied troops to German panzer groups.[267]

By the following month, Hitler's army occupied Denmark, along with Norway which held special strategic importance through access to sea routes that supplied crucial resource supplies for the Nazi war machine. Sweden remained the only Scandinavian country to successfully maintain neutrality throughout the war.[268]

France, which was then considered to have the world's best army, fell to Nazi occupation after one month of attacks when Marshal Philippe Pétain surrendered. Hitler allowed Pétain to remain as dictator of an area known as Vichy France on the Atlantic coast. The French general was tried for treason at the end of the war. His death sentence was later commuted to life imprisonment.

France's military debacle led to Hitler's greatest strategic miscalculation which later turned the tide of the war. Hitler refused to risk committing his panzers to action at the French seaport of Dunkirk where more than 300,000 British and French troops were trapped. While braving attacks by German Luftwaffe aircraft, British warships along with a volunteer armada of large and small private vessels successfully evacuated an estimated 198,000 British and 140,000 French soldiers to England.

Some members of the escaped French troops formed around General Charles de Gaulle's Free French forces which continued to battle Hitler. Many escapees also later formed the core of army troops that recaptured Normandy beaches in northern France on D-Day (June 6, 1944).

During the summer of 1940, Hitler launched massive Luftwaffe bombing attacks against the British Isles (the Battle of Britain). The terror directed against civilian targets included the V-1 flying bomb—also known to the Allies as the buzz bomb— the first of Germany's so-called vengeance weapons designed to demoralize London citizens.

While first suffering great devastation, Great Britain's Royal Air Force eventually turned the air war against the aggressors, shooting down 2,698 German planes while losing only 915. This reversal marked the first of Hitler's major defeats.[269]

Despite having signed a nonaggression pact with Stalin, Hitler redirected his territorial ambitions eastward to the Soviet Union. More than three million German troops attacked the USSR on June 22, 1941—the largest invasion force the world had ever seen. Although he had been warned, both by other countries and by his own intelligence network, Stalin had refused to believe it. Unprepared, and initially suffering early setbacks, Stalin ordered counterattacks which stalled the German advance at the gates of Moscow.[270]

All-in-all, Mussolini's Italy had relatively little impact on World War II. After Il Duce, as he was called, had launched an unsuccessful North Africa offensive from Italian-controlled Libya into British-controlled Egypt, Hitler came to his aid by sending a few thousand troops, a Luftwaffe division which recaptured the strategic port city of Tobruk, Libya. Nazi support forced the British Eighth Army, which had been positioned to capture Libya, to retreat back into Egypt.

In spring of 1942, Hitler launched a fresh Soviet Union offensive aimed at capturing the oil-rich Caucasus and the city of Stalingrad. The German 6th Army ultimately ran out of ammunition and rations and surrendered, dealing a severe blow to Hitler's Eastern ambitions.

The United States declared war on the Japanese Empire four days after Japan's December 7, 1941, attack on Pearl Harbor in Hawaii. On the same day, December 11, Germany declared war against the United States, and the United States also declared war on Germany.

Germany had been obliged by a Tripartite Pact agreement to come to Japan's aid if the country was attacked by another country, but not if Japan was the attacker. Hitler appeared to dismiss the distinction between aggrieved and aggressor and declared war anyway. Italy immediately joined Germany in declaring war against America.

Much to the chagrin of Stalin, the U.S. and British actions against Germans were relatively limited from 1942 to 1944. The Soviets had begun to gain ground during this period following some victories such as a tank battle against General Erwin Rommel's panzers at Kursk and Germany's final strategic offensive on the Eastern Front. An Allied invasion of Sicily on September 9, 1943, had forced Hitler to redirect a large body of his forces to counter the Mediterranean challengers.[271]

Mounting Germany's extensive losses of troops and tanks ensured that the victorious Soviet Red Army would enjoy a strategic advantage throughout the remainder of the war. Rommel had been forced by 1943 to abandon North Africa after a defeat by Montgomery at El Alamein in what was to be the first decisive Allied victory over the German army in what Churchill declared was "the end of the beginning." [272]

By the beginning of 1944, Hitler had lost all initiative against the USSR, and was struggling to hold back the Allied tide on the European Western Front. By the winter of 1943, the southern half of Italy was in Allied hands, Mussolini was stripped of power, imprisoned and later executed by his people on April 28, 1945. By June 4, 1944, Rome had fallen to the Allies.

The Battle of the Atlantic occurring from 1942 to the war's end in 1945 which has been described as "the largest and most complex naval battle in history" was initially waged to counter German U-Boat actions to sever vital shipping supply lines between Britain and America. U.S. destroyer battleships and aircraft with extended patrol ranges were crucial developments that essentially ended the U-Boat shipping threat by late 1943.

The Allied June 6, 1944, D-Day offensive codenamed Operation Overlord on five Normandy beaches along a 50-mile stretch of the heavily-fortified French coast was a significant development leading to release of Western Europe from Nazi control. Although the 1,200 airborne assault in support of amphibious 156,000 American, British and Canadian landing forces gained only a tenuous foothold on the first day, their numbers expanded to more than two million Allied troops by the end of August.

Allies succeeded in capturing Cherbourg, France, on June 26, and Caen on July 21. They also beat back an August 8 counterattack which left 50,000 of Germany's defeated 7th Army trapped and unable to escape.

A second Allied invasion of southern France from the Mediterranean Sea (Operation Dragon) on August 15th led to the liberation of Paris five days later.

How Everything Happened, Including Us

In the winter of 1944, Hitler made a last desperate and failed war gamble in what is known as the Battle of the Bulge. The introduction of new, more modern Allied tanks and disadvantage of dwindling German troop numbers were decisive influences. Nevertheless, the casualties were terrible on both sides, making it one of the bloodiest battles of the war. It was also one of the costliest in all of American Army history.

By this time, the war was looking ever darker for Hitler. Soviets forces had advanced nearly to the Polish-Soviet border. On July 20, 1944, a group of conspiring German officials unsuccessfully attempted to assassinate the Fuhrer (leader).

Adolf Hitler took his own life on April 30, 1945. One week later, on May 8, the Allies formally accepted the unconditional surrender of Germany.

Impressive advances of new military weapons had arrived too late to change the tide of war. Included was the replacement of the V-1 flying bomb with a faster V-2 flying bomb providing a larger payload jet aircraft which was vastly superior to propeller models, and submarine improvements which might have changed decisive outcomes of many Atlantic naval battles.

In early February 1945, three Allied leaders: Franklin Roosevelt, Winston Churchill and Joseph Stalin, met in newly-liberated Yalta on the Crimea of the Soviet Union to divide up territories of post-war Europe. Most of the East went to Stalin who disingenuously agreed to allow free elections. The West went to Britain, France and the United States. Post-war Germany was split up between the four, as was the city of Berlin.

Agreements at the Yalta Conference (also known as the Crimea Conference) established divisional spheres of international influence which continue to dominate international diplomacy. Rivalries that began even before that war, combined with actions of victorious powers, laid foundations for an Iron Curtain in Europe, and East-West Cold War tensions which still remain evident.

Larry Bell

World War II in the Asia-Pacific Theater

The World War II Asia-Pacific Theater, often referred to as the Pacific Theater, pitted Allied Western powers (including Australia, the United States and Britain) against Japan. This theater of operations covered a vast area with four main areas of conflict: the Pacific Ocean and islands, the South West Pacific and South East Asia and China.

Other Allied Pacific Theater participants included China and the armed forces of British India, New Zealand, Canada and the Netherlands (which possessed the Dutch East Indies and the western part of New Guinea) who were partner members of a Pacific War Council.

Tensions leading up to the Pacific Theater conflict had been building from as early as the mid-1930s over Japanese military intentions to take control over Western-dominated Dutch East Indies oil reserves. In their quest, the Japanese had targeted Indochina, Malaya and the Philippines for purposes of creating a Greater East Asia Co-Prosperity Sphere.

The Western powers responded to what they recognized to be a military threat by denying Japanese sales of oil, iron ore and steel which would support its expansionist ambitions. Viewing the embargos as acts of aggression, the Japanese media, influenced by military officials, pushed for retaliatory actions. The Japanese Imperial General Headquarters had begun planning for a war with the Western powers by 1941.

The Japanese Empire's plan to control of East Asia began with a 1931 event commonly known as the Manchurian Incident in which the Japan military staged a bombing of the South Manchurian Railway which it attempted to blame on Chinese dissidents as a pretext for invading northeast China. The ruse was exposed and denounced by the League of Nations which subsequently expelled and internationally isolated Japan. Nevertheless, Japan had succeeded in turning that region of

Manchuria into a puppet-state constitutional-monarchy controlled by Japan.

Strained relations with China over Japan's occupation of Manchuria led in 1937 to a bloody Second Sino-Japanese War. Although Japan enjoyed some early major victories, capturing both Shanghai and the Chinese capital of Nanjing, the war later reached a stalemate. In September 1940, unable to control the large Chinese cities and vast countryside, Japan joined the Tripartite Pact with Germany and Italy.[273]

Japanese forces had also attempted to advance through Manchuria into the Soviet Far East but were defeated by a mixed Soviet and Mongolian force in a 1938 Battle of Khalkhin Goi. Earlier Soviet aid to China had ended upon the signing of a Soviet-Japanese Neutrality Pact at the beginning of its war against Germany.[274]

The Soviet Union had fought two brief undeclared border conflicts with Japan in 1938 and 1939, and then remained neutral until August 1945 when it joined with Allies and together invaded the territory of Manchukuo, China, Inner Mongolia, the Japanese protectorate of Korea and other regional territories.

In September 1940, Japanese forces cut China's only landline to the outside world by seizing Indochina which was then-controlled by Vichy France headed by Marshal Philippe Pétain. After then breaking their agreement with the Vichy administration, the fighting which resulted, ended in Japanese victory.

On September 27, 1940, Japan signed a military alliance with Germany and Italy, becoming one of the three Axis powers. However, there was actually little World War II coordination between Japan and Germany until 1944. Other than operating submarines and raiding ships in the Indian and Pacific Oceans and sharing Japanese naval facilities, German and Italian involvement in the Pacific War was overall quite limited.

A gamble that their surprise early morning December 7,

1941, Japanese air strike on the U.S. Pacific Fleet at Pearl Harbor, Hawaii, would persuade America to negotiate a settlement which would cede free rein to Japan in Asia turned out to be a very bad wager. Instead, the massive attack drew an opposite response.[275]

The Pearl Harbor attack which killed 2,404 military and civilians, took eight battleships out of action, and destroyed 188 aircraft shocked America into fully-committed retaliatory action.

As some consolation to the event, the devastation might have been even worse. Locally-based American aircraft carriers were out at sea when the attack occurred. The bombing and strafing had also left vulnerable naval infrastructure (such as fuel oil storage, shipyard facilities and a power station) and a submarine base unscathed.

The Japanese attacked Thailand and British Malaya on the same day as Pearl Harbor, with their forces advancing towards a large British naval base in Singapore. Malaya and Singapore fell to Japanese control by February 1942.

The Imperial Japanese Army established a major base in Australian territory of New Guinea in early 1942, and by 1943 had air-attacked the Australian mainland about 100 times. An estimated 15,000 Australian soldiers were taken as prisoners, and by the war's end, 22,000 Australians were captured by the Japanese; 8,000 died as war prisoners.

In addition to defending against Japan, Australia was also committed to fight Hitler in the Mediterranean Sea. Lacking vital armaments, modern aircraft, heavy bombers and aircraft carriers, Australian Prime Minister John Curtin had called upon American support.

Triggered by Pearl Harbor, in March 1942 U.S. President Franklin Roosevelt ordered General Douglas MacArthur to formulate a Pacific defense plan with Australia. MacArthur became Southwest Pacific Supreme Commander, established his headquarters in Melbourne and began amassing U.S. troops there.

Undaunted by the American presence, enemy naval activity proceeded to Sydney. In late May 1942, Japanese midget submarines launched a raid on Sydney Harbor and shelled eastern suburbs and the city of Newcastle.

U.S. actions were concentrated on two war theaters: The Pacific Theater and the China Burma India (CBI) Theater. Allied operational control was divided between two supreme commands: The Pacific Ocean Areas and the Southwest Pacific area.

The first of several major Allied offensives was the May 1942 Battle of the Coral Sea between U.S. and Australian naval and air forces versus the Japan Imperial Navy ended with mixed results for both sides. Whereas the Japanese scored a tactical victory in terms of numbers of ships sunk, the Allies gained an important longer-term strategic advantage in checking the Japanese advance.

The Allies exploited Japan's strategic vulnerability two months later in the South Pacific and launched Guadalcanal and New Guinea Campaigns which further broke down their defenses.

The June Battle of Midway, which commenced early in the following month (six months after the Pearl Harbor attack), inflicted irreparable damage to the Japanese fleet. Military historian John Keegan described the Allied victory as "the most stunning and decisive blow in the history of naval warfare."

Under commands of U.S. Admirals Chester Nimitz, Frank Jack Fletcher and Raymond A. Spruance, Midway was the last great naval battle for two years. Meanwhile, the United States turned up its vast industrial potential to increase its numbers of ships and aircraft.

Allies began a long movement across the Pacific, seizing one island after another. The goal was to get close to Japan, launch massive air attacks, improve a submarine blockade and potentially (but only if necessary) execute a land invasion.

The first major Allied forces offensive against the Japanese Empire was the Battle of Guadalcanal launched primarily by U.S. Marines between August 1942 and February 1943. Supported by American and naval forces, the successful landing assault on Guadalcanal, Tulagi and Florida in the Solomon Island was staged to deny Japan their use to threaten Allied supply and communication routes between the United States, Australia and New Zealand.

Several months later, Japanese attempts between August and November 1943 to retake a vital airfield culminated in a crushing naval battle defeat for Japan which marked a turning point for Allies to take the offensive in Solomon Islands, New Guinea and the Central Pacific.

America suffered heavy losses in troops, cruisers and carriers in gaining that victory. However, the constant pressure to reinforce Guadalcanal had weakened Japanese efforts in other theaters and contributed to a successful Australian and American counteroffensive in New Guinea which culminated in the capture of key Japanese offensive bases on Buna and Gona and freed Australia from future invasion threats.

Remaining Japanese defenses in the South Pacific area were then either destroyed or bypassed by Allied forces as the war progressed.

One of the bloodiest American Pacific War battles was fought over the small eight-square-mile island of Iwo Jima situated halfway between Tokyo and the Mariana Islands. Allies aimed to capture the island and prevent its use as an early-warning station against air raids on the Japanese Home Islands, and also to secure its use as an emergency landing field.

Japan's commander of Iwo Jima defense, Lt. General Tadamichi Kuribayashi, fully recognized that his troops could not win the battle. By 1944 the Imperial Japanese Navy had lost nearly all of its defensive power to prevent the landings, and the maximum 550-mile range of the Empire's remaining aircraft could not reach Iwo Jima from home bases.

Nevertheless, Kuribayashi had planned to make the Americans suffer more than they could endure.

By the time of America's February 19, 1945 invasion, the Japanese had dug an extensive network of deep defensive tunnels, command centers and barracks. In addition, hundreds of hidden artillery and mortar positions (including explosive rockets) along with land mines were placed throughout the island.

Still, despite heavy casualties, America's troops ultimately prevailed. On February 23, 1945, the 28th Marine Regiment reached the summit of volcanic Mt. Suribachi, prompting the now famous Raising the Flag on Iwo Jima picture. The flag raising is often cited as the most reproduced photograph of all time and became the archetypal representation not only of the Battle of Iwo Jima, but of the entire Pacific War.

For the rest of February, the Americans pushed north, and by March 1, 1945, had taken two-thirds of Iwo Jima. By March 26, the island was finally secured.

The Japanese fought to the last man, killing 6,800 Marines, and wounding nearly 20,000 more. Japanese losses were even greater, totaling well more than 20,000 men killed.

Altogether, hard-fought battles on the Japanese home islands of Iwo Jima, Okinawa, and others resulted in horrific casualties on both sides. Of the 117,000 Okinawan and Japanese troops defending Okinawa, estimated 94 percent died.[276]

As the war progressed, the Japanese military turned to increasingly desperate measures. Faced with the loss of most of their experienced pilots, the Japanese expanded their use of kamikaze suicide tactics in an attempt to create unacceptably high Allied casualties. Allies responded with a naval blockade of Japan's mainland in combination with air raids.

On August 6, 1945, the United States dropped an atomic bomb on the Japanese city of Hiroshima. Following the bombing, President Harry Truman issued a press release warning the

Japanese either to surrender "…or expect a rain of ruin from the air, the like of which has never been seen on this Earth."

Truman didn't exaggerate. Three days later, the United States dropped a second atom bomb on Nagasaki. Between 140,000 and 240,000 citizens of the two cities perished.

On August 9, 1945, the Soviet Union declared war on Manchuria and invaded that country. Together with the bombings of Hiroshima and Nagasaki, the effects of these "twin shocks" prompted the Imperial Japanese Cabinet to issue a "sacred decision" to accept terms of surrender provided that Japan would retain the "prerogative of His Majesty [the Emperor] as a Sovereign Ruler."

Following an agreement negotiated among Allied leaders at a Potsdam Conference in Germany, the American government responded on August 15 that the authority of the Emperor "shall be subject to the Supreme Commander of the Allied Powers." The Potsdam representatives included U.S. President Harry Truman, British Prime Minister Winston Churchill, Anthony Eden (plus Clement Attlee and Ernest Bevin who succeeded the pair during the conference) and Soviet Premier Joseph Stalin.

Although the Pacific War ended on August 14, 1945, Imperial Japan announced its intent to surrender on August 15, a date known in English-speaking countries as V-J Day (Victory in Japan day). The formal Japanese Instrument of Surrender was signed on September 2, 1946 aboard the USS Missouri battleship in Tokyo Bay. Following this time, MacArthur went to Tokyo to oversee Japan's post-war development.

Although a Shinto Directive forced Japan's Emperor to relinquish much of his authority and his divine status, it paved the way for extensive and lasting extensive cultural and political change.

In the aftermath of the war, Japan lost all rights and titles to its former possessions in Asia and the Pacific, and its sovereignty was limited to the four main home islands at enormous costs to both sides.

America alone suffered an estimated 426,000 human casualties: 161,000 dead (including 111,914 in battle and 49,000 non-battle) and 16,358 captured (not counting POWs who died). The United States also lost 21,355 aircraft, along with nearly 200 warships, including 5 battleships, 11 aircraft carriers, 25 cruisers, 84 destroyers and destroyer escorts and 63 submarines.[277]

World War II dramatically changed international identities and intergovernmental political alignments.

Much of Western Europe, which lay in ruins from aerial, naval and land bombardments, was rebuilt with assistance from the Marshall Plan. Germany was placed under joint military occupation for the next seven years by the United States, Britain, France and the Soviet Union, and Berlin, although in Soviet-controlled territory, was divided among all four powers.

Remarkably, despite military occupation, Japan and Germany rapidly rose to become and remain two of the world's most powerful economies.

Almost all of the major nations that were involved in World War II soon began shedding their overseas colonies. European powers began a decades-long process of withdrawing from their possessions in Africa and Asia. Indian Independence ended British Rule in 1947 with territorial partitioning between modern-day India and Pakistan. This later led to the creation of the People's Republic of Bangladesh in 1971.[278, 279]

Armed insurrections in Indochina drove the French out in the early 1950s and led to the formation of Laos, Cambodia and Vietnam. Between 1956 and 1962, nearly 20 African countries also achieved their independence from France, and African nationalists led Kenya and Ghana to independence from foreign rule.[280]

The emergence of newly independent countries in Africa and Asia often created conflicts between historically antagonistic ethnic or religious groups who either came to share the same country as well as between multiple nations with territorial claims over regions which had previously been divided differently. The bloodiest of these included the Nigerian Civil War, the Second Congo War, the Second Sudanese Civil War and the Bangladesh Liberation War.

Failure of the League of Nations to prevent World War II led to its dissolution. It was replaced by a new United Nations organization on October 24, 1945, in a renewed attempt to maintain world peace. By 1946, there were 35 UN member states. Joined by newly independent nations, that membership grew in number to 127 by 1970.

The 1945 Yalta Conference agreement divided Western capitalist powers and the communist Soviet Union into separate European spheres of influence which set the stage for a geopolitical rivalry that would come to dominate international relations.

A Soviet iron curtain descended across the continent from Stettin in the Baltic, to Trieste in the Adriatic. Behind that line were located all capitals of the ancient states of Central and Eastern Europe: Warsaw, Berlin, Prague, Vienna, Budapest, Belgrade, Bucharest and Sofia.

Over the course of the war, the Soviet Union had already annexed several countries as Soviet Socialist Republics (SSRs). Eastern Poland was incorporated into Belarusian and Ukrainian SSRs; Latvia, Estonia and Lithuania became SSRs; part of eastern Finland became a Karelo-Finnish SSR; and eastern Romania became a Moldavian SSR.

Then, between 1945 and 1949, Yugoslavia, Albania, Bulgaria, Poland Romania, Czechoslovakia, Hungary and East Germany became independent communist People's Republics with close Soviet ties as de facto satellite states.[281, 282]

The rise of communism also spread outside Europe, adding nations of Mongolia, China, North Korea and Vietnam. This expansion of communist ideology and Soviet influence created a deep and lasting rift between many former World War II allies. The two emerging rival blocks coalesced into formal competing mutual defense organizations, forming the North Atlantic Treaty Organization (NATO) in 1949 and the Warsaw Pact among the USSR and its seven satellite states of Central and Eastern Europe in 1955.[283]

World War II very substantially reshaped global power structures in another way: the introduction of incredibly powerful and horrifically devastating nuclear warfare capabilities. Fearing that Nazi Germany was pursuing the development of these weapons, the United States and the United Kingdom pooled efforts to beat them to it. A then-top-secret Los Alamos, New Mexico Manhattan Project developed two such devices: the uranium-235-fueled Little Boy dropped on Hiroshima and the plutonium-239-powered Fat Man that struck Nagasaki.

As discovered after the war, the German nuclear program had lagged far behind.

Meanwhile, with post-war Western relations rapidly deteriorating, the Soviet Union, supported by espionage efforts, developed and detonated its first nuclear weapon in August 1949. The United States countered through a crash program to create the first hydrogen bomb in 1950 and detonated an even more destructive second-generation thermonuclear weapon in 1953 which was more than 400 times as powerful as those dropped on Japan. The Soviet Union then followed suit, detonating a primitive thermonuclear weapon in 1953, and a full-fledged version in 1955.

Development of computerized long-range nuclear delivery systems by both camps produced a rapidly accelerating and increasingly dangerous mutually-assured-destruction Soviet versus United States arms race. Tensions led to a broader proliferation of nuclear weapon development and stockpiling. Several other nations, including the United Kingdom, France, China, India, Pakistan, North Korea and Israel are believed to

Larry Bell

have gained first-strike and retaliatory capabilities.[284]

Wars by Proxy: Korea and Vietnam

Capitalist versus communist conflicts during the 1950s provoked two major war outbreaks, one primarily between the China-backed North Korea and United States-backed South Korea; and a second involving the United States in North Vietnam.

The first of these wars was fought over total sovereign control of the Korean Peninsula, a former Japanese colony which was divided along the 39th parallel into a communist territory north of the border, and a capitalist-governed region to the south. Intense fighting broke out on June 25, 1950, following a series of border clashes when North Korea, under the leadership of Kim Il Sung, supported by the Soviet Union and China, invaded South Korea.

Led by the United States, which provided about 90 percent of the military personnel, 21 United Nations went to the South's defense. Facing near-defeat after the first two months, they launched a counter-offensive which effectively cut off North Korean opposition at Incheon. In October 1950, the advancing UN forces were confronted by a massive surprise Chinese intervention at the Yalu River border with China. The UN retreat continued until mid-1951.

Throughout repeated battle reversals, Seoul changed hands. As the war waged on, North Korea was subjected to a massive bombing and air-to-air combat campaign. Soviet pilots covertly engaged American and allied pilots in North Korea's defense.

Although an armistice signed on July 27, 1953 created a Korean Demilitarized Zone that separated the North and South, no peace treaty between these two independent, sovereign nations was consummated.

The Vietnam War occurred following French withdrawal from that former colony on July 21, 1954. As with Korea,

Vietnam was divided into two halves, a communist North and capitalist South, this time along the 17th parallel. Border conflicts between the two sides again escalated into a regional civil war.

Although initially providing only material aid to the South, the United States didn't become militarily engaged in fighting until a Gulf of Tonkin Resolution was passed in reaction to two alleged North Vietnamese attacks on American naval destroyers.

The original American report blamed North Vietnam for both incidents, but eventually became very controversial with the widespread belief that at least one, and possibly both incidents were false, and possibly deliberately so.

In the first August 2, 1964 incident, the destroyer USS Maddox was reportedly pursued and exchanged fire with three North Vietnamese Navy torpedo boats while on an intelligence patrol. Four North Vietnamese sailors were killed along with six wounded. No U.S. casualties resulted, and the Maddox was hit by only a single machine gun round.[285]

Another National Security Agency report claimed that a second Gulf of Tonkin incident occurred two days later, a representation that U.S. Secretary of Defense Robert S. McNamara later admitted never actually happened.[286]

Originally premised upon a "domino theory" fight to contain communism, non-voluntary drafting of troops and disturbing news coverage about a devastating Tet Offensive and scandalous My Lai massacre turned popular U.S. sentiments against American engagement in the war. As South Vietnamese forces were pushed back, hostilities expanded into neighboring Cambodia.[287]

Ultimately, the United States and North Vietnam signed Paris Peace Accords which ended America's involvement in the war. However, with the threat of U.S. retaliation over, the North proceeded to violate the ceasefire and invaded the South with full military force. Saigon was captured on April 30, 1975, and Vietnam was unified, this time under communist rule, the following year.[288]

The Near-War Cuban Missile Crisis

Major post World War II communist-capitalist territorial power reshuffles and conflicts in combination with terrifying nuclear weaponry and long-range rocket delivery systems led the Cold War world ever-closer to the brink of mutually-assured-destruction that such developments were at least theoretically intended to prevent. Holocaust was narrowly averted in the aftermath of a tense 13-day October 1962 political and military standoff between U.S. President John Kennedy and Soviet leader Nikita Khrushchev over missiles being shipped to Cuba.

The Cuban Missile Crisis occurred after Lockheed U-2 spy planes revealed missile launchers being installed over the U.S. neighbor island which was controlled under Fidel Castro's socialist government with close Soviet Union ties. In response, Kennedy instituted a naval blockade around Cuba to block Soviet missile shipments.

Threatening to penetrate the defense, military conflict was avoided when the USSR backed down and agreed to remove the missiles in exchange for a U.S. commitment not to invade Cuba.

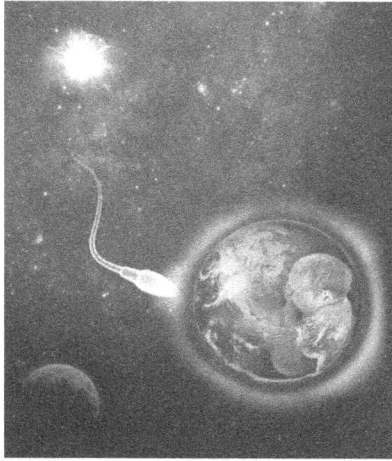

Part Eleven: Pioneering a New Space Age

U.S. LEADERSHIP AND citizens had become highly concerned about Soviet dominance in advanced missile technology and its military applications since October 4, 1957, when a satellite named Sputnik chirped a wakeup call from Earth's orbit.

America's psyche was again jolted by more shock waves three and one-half years later on April 12, 1961, when a young Russian cosmonaut named Yuri Gagarin leant his human face to a new extraterrestrial space era that threatened to leave the United States behind.

That wasn't the only humiliation facing President John Kennedy's administration at the time. A Cuban Bay of Pigs debacle had just occurred in mid-April as well.

The particularly unfortunate timing of these two events put great pressure on Kennedy to demonstrate resolute leadership. On May 25, 1961, only a few weeks after Gagarin's orbital flight, he upped the ante, committing the United States to send a man to the Moon and to return him safely before the end of that decade.

Kennedy rallied the country to that cause, saying:

> ...*no single space project in this period will be more impressive to mankind, or more important for the long-range exploration of space; and none will be so difficult or expensive to accomplish...in a very real sense, it will not be one man going to the Moon—if we make this judgment affirmatively, it will be an entire nation. For all of us must work to put him there.*

The president clarified that this was to be a competitive race dedicated to demonstrating U.S. technical supremacy over the Russians. He said:

> *Within these last 19 months at least 45 satellites have circled the Earth. Some 40 of them were made in the United States of America and they were far more sophisticated and supplied far more knowledge to the people of the world than those of the Soviet Union.*

America exceeded Kennedy's decadal commitment, putting four of our citizens on the lunar surface and returning them by 1969, plus delivering two more into lunar orbit who returned with them.

Within three more years, eight other NASA astronauts had walked on the Moon on successful round-trip voyages, along with four more orbital companions. Some of those same astronauts literally blazed that pathway. They flew on two suborbital and four Earth-orbital Mercury launches, nine Earth-orbital Gemini flights, two Earth-orbital Apollo tests and two lunar-orbital tests that made those lunar surface landings possible.

How Everything Happened, Including Us

Many of those early space pioneers were and are my personal friends. For example, Neil Armstrong served on the board of a commercial aerospace company I co-founded. Buzz Aldrin is a close long-time friend, professional colleague and frequent home guest. Apollo Seven Lunar Module Pilot Walter Cunningham serves on the advisory board of the Sasakawa International Center for Space Architecture I founded at the University of Houston.

The original technology that literally gave rise to the space race and subsequent military and peaceful applications can be traced back to a dark pre-World War II German legacy. In 1934, a young physicist named Wernher von Braun who had just received a doctorate degree from the University of Berlin had succeeded, with members of his academic group, in launching two rockets that reached between one and two-mile altitudes. His graduate rocketry research was conducted at a solid-fuel rocket station not far from Berlin under the supervision of then-Captain Walter Dornberger, a department head for the German armed forces Ordinance Department.

During this time period, the National German Workers Party (NSDAP, or Nazi party) came into power and moved rocketry into the national agenda. Von Braun's thesis titled *Construction, Theoretical, and Experimental Solution to the Problem of Liquid Propellant Rocket* was kept classified by the German government until 1969.

In the early 1940s, von Braun moved to the new Peenemunde facility as its technical director under the command of Captain Walter Dornberger where his group, in combination with the Luftwaffe, developed liquid-fuel rocket engines for aircraft and jet-assisted takeoffs. Even more significantly, Peenemunde became the development center for a new A-4 ballistic missile which was to become better known as the previously-discussed V-2.

Adolf Hitler recognized the importance of using the A-4/V-

2 for military purposes soon after Germany invaded Poland to start World War II in 1939. Von Braun, who was imprisoned on espionage charges for resisting an attempt by Gestapo Chief Heinrich Himmler to take control of the V-2 project, was reportedly released under Hitler's personal order.[289]

Space Race Competition for German Technology

Although the German V-2 program was not generally regarded to have been a decisive factor in the WWII outcome, the remarkable new rocket weapon caught the keen attention of the USSR and America's allies. British Prime Minister Winston Churchill requested Joseph Stalin's permission for English specialists to go into Poland in order to investigate the V-2 test range located in the region of Soviet attack forces.[290]

The Soviets already knew at that time about the German jet-powered V-1 buzz bomb, but it was judged not to be powerful enough to represent great concern. On the other hand, their own investigators in Poland had recovered fragments of a much more advanced A-4/V-2 which the Germans had unsuccessfully attempted to completely destroy. Close inspection of those remnants caused major alarm. Soviet specialists were astounded to calculate that the rocket engine was capable of lifting at least 20 tons.

The Russians were also shocked to discover that the German rocket was fueled by alcohol and liquid oxygen technology rather than by the customary nitric or kerosene they used. With a gross weight of 12.52 tons, the V-2's advanced turbopump engine design, powered by an 80 percent hydrogen peroxide steam generator, could deliver a 1-metric-ton payload over a distance of 200 kilometers.[291]

Even during the war, the United States and Soviet Union competed to gain rocket technology secrets from German engineering leaders and specialists. America scored a great advantage through Operation Paperclip by offering citizenships and employment to von Braun and his entire Peenemunde team. According to the Russian newspaper *Izvestia*, the Soviets were so interested in knowing what Operation Paperclip was about that they placed agents in the U.S. zone with plans to capture Von Braun. That never happened because he was too closely guarded.[292]

Meanwhile, the USSR undertook concerted efforts to round up as many remaining V-2 technicians, blueprints and as much scavenged hardware as could be found in East Germany's Soviet zone. V-2 reconstruction and analysis work was undertaken in a shabby three-story East German electric power station located in Bleicherode near Nordhausen. The activities eventually involved about 1,000 people: approximately half Russian and half German workers, plus between 50-60 Peenemunde technical veterans.[293]

By October 1945, the American von Braun team was launching a series of V-2 demonstrations. Organized by the British near Cuxhaven on the North Sea, the firings were expedited in recognition of the fact that Nordhausen, where the rockets were manufactured, would soon be in the Russian zone.

The team scavenged enough weapon material and launch apparatus to fill 200 trucks and 400 freight cars, but ultimately managed to assemble only eight complete missiles and launch three. These operations involved approximately 2,500 British military personnel, along with about 1,000 German nationals including 25 veteran Peenemunde specialists and 274 former war prisoners.[294]

In addition to shipping enormous quantities of German V-2 equipment items to Russia, the Soviets transported as many as 5,000 skilled German technicians and their families there as well...and not always on a voluntary basis. Those included were not all rocket specialists, but also engineers, scientists and technicians experienced with weapons systems, aircraft and submarine development.

A freshly released Soviet prisoner still technically under conviction for counter-revolutionary activity believed that the V-2 was already obsolete. Publicly known only as Chief Designer, his top-secret-protected name was Sergei Pavlovich Korolev.

Under Korolev's leadership, the Russians began to produce their own designs for a new R-1 program. Several technologies were developed from scratch at 35 research stations and design bureaus along with 16 principal plants. By 1938 they had built 12 R-1s, of which seven of nine launched hit their targets.[295]

Within ten years, Korolev's group had created a series of rockets progressing from the R-1 first commissioned in 1948 to the R-7, the USSR's—and the world's—first intercontinental ballistic missile (ICBM) which is still in service. Its 600-kilometer range was twice that of the V-2/R-1, plus provided a warhead separation capacity to propel that payload rather than the entire vehicle to a target.

Beginning in January 1952, Soviet-hired German rocket technicians were rapidly being returned to their Fatherland. By this time the Russians, who had milked them of their expertise and technology, were well on their way to higher program trajectories without the need of further German assistance.

A new Space Age had been launched.

Early Dreamers, Designers and Developments

The space age was founded and shaped upon bold ideas and dedication of many great minds whose contributions of purpose, passion, professionalism and persistence have brought humankind to our present crossroads of great possibilities and uncertainties. Four among countless others of these visionaries include a remarkably innovative Russian school teacher, another Russian who survived terrible deprivations in a prison work camp, a German World War II rocket developer and an American who dared to believe that rockets can operate in a space vacuum.

Their combined story, and those who joined and followed them, is one of historic achievements, events and lessons born of triumphs and tragedies. It reveals a nexus of politically-manipulated and ideologically-shifting public rivalries between nationalistic pride and paranoia where space exploration and technology manifests full dimensions of civilizations' boldest dreams and greatest fears. In all cases, it has inexorably changed and expanded our world.

Konstantin Tsiolkovsky's Remarkable Vision

As historian Walter A. McDougal observed in his 1985 book *Heavens and the Earth*, there is probably no more exemplary and ironic time and place to begin this narrative saga than in early Bolshevik Russia:

> *Modern rocketry and social revolution grew up together in tsarist Russia. There is no anomaly in the fact that the most 'backward' of the Great Powers before World War I was the one that fostered violent rebellion against the chains of human authority and the chains of nature.*

Russian thinkers dating back to the 1880s including Viktor Sokolsky have contemplated general possibilities of creating the liquid-fueled rockets which are commonplace today. We can thank the writings of a self-taught, high school mathematics and physics teacher in the small town of Kaluga south of Moscow for the concepts and calculations upon which such realities depend.

Broadly considered to be the Father of Space Travel, Konstantin Eduardovich Tsiolkovsky (1857-1935) originally preoccupied his early years with personal design studies related to research into stellar radiation and design concepts for steam engines and metal-fabricated dirigibles. Then, upon conceptualizing possibilities for reaction-driven devices later called rockets, he published an amazing book in 1883 titled *Free Space* which proposed a comprehensive, detailed and ingenious design for a liquid-fueled propulsion device for use in the vacuum of space...

> *...which, when combined chemically would yield per unit mass of resultant product such an enormous amount of energy.*[296]

Konstantin Tsiolkovsky's prolific productivity during the 1920s through early 30s was amazing, conceiving ideas for a reaction engine (1927-28); a new airplane (1928); a jet-propelled aeroplane (1929); the theory of the jet engine (1930-34); the maximum speed of a rocket (1931-33); and the final classic work before his death, space rocket trains (1924-1934).

He told a group of students at the Zhukovsky Academy in 1934:

> *...I am not at all sure, of course, that my 'space rocket train' will be appreciated and accepted readily, at this time. For it is a new conception reaching far beyond the present ability of man to make such things. However, time ripens everything; therefore I am hopeful that some of you will see a space train in action.*

And they did. For example, Tsiolkovsky providently conceived space rocket trains which are now the standard multi-staging technique used to deliver payload elements to Earth orbits and planets. His enormous conceptual achievements led to the USSR's first Mir orbital space station which was realized in large part through the efforts of another important Russian designer in 1971.

Russia's Secret Hero, Sergei Pavlovich Korolev

Tsiolkovsky's rich legacy of design contributions guided design practices of a great, officially unnamed Soviet engineer known by his colleagues as "SP" who led efforts which produced the USSR's first ICBM and launched the Space Age with the first orbiting satellite (Sputnik), the first dog, first man, first two men, first woman, first three men, first spy, communication satellites and the vehicles and spacecraft that first reached the Moon and Venus and passed by Mars and the Mir orbital station.

The Chief Designer's identity was concealed as a state security secret under orders from Stalin, Khrushchev and Brezhnev until his death in 1966 during the peak period of a USSR race to beat America in landing its citizens on the Moon. Only then was he publicly honored as the Hero of Socialist Labour.

Sergei Pavlovich Korolev's remarkable career began with a childhood passion for aviation.

By 1929 he developed his own unpowered glider. This primary flying interest soon turned to possibilities of using rockets to improve aircraft performance.

While still a young university student, Korolev joined a Group for Studying Rocket Propulsion (GIRD) and began working on a small gasoline-fueled propulsion engine-powered rocket weighing 40 pounds. Launched in August 1933, it flew for a total of only 18 seconds. Nevertheless, he enthusiastically predicted in an article titled *Towards the Rocketplane:*

> *Jet flight vehicles can develop flight speeds of 3,600 km/hr...and [can attain] immense altitudes [but that] practical resolution of this huge problem requires years of persistent work.*

GIRD activities by Korolev and his coworkers drew the attention of the Russian military. His rocket group became merged into a new government-headed organization called the Reaction Propulsion Institute (RNII) for key purposes of creating reliable and accurate rocket guidance and control systems.

A massive purge of USSR scientists, engineers and military leaders accused of trumped-up charges of spying for Germans led to the arrests of many RINN engineers and the execution of the organization's main sponsor, renowned military hero Marshal Tukhachevsky. On the early morning of June 27, 1938, two KGB agents took then-31-year-old Korolev into custody with no time to say goodbye to his three-year-old daughter Natasha.[297]

Having risen to a high-level RNII position, Korolev was taken to Lefortovo prison where he was interrogated and beaten. Upon asking for a glass of water he was hit on the head by a jug and called an enemy of the people. He was then told, "Today is your trial" and was led down a long corridor into a room.

When the door opened, Kliment Voroshilov, one of Stalin's closest associates, entered. Imagining that Voroshilov would straighten out the problem Korolev told him "I didn't commit any crime." Voroshilov then shouted, "None of you swine ["svolochi" in Russian] have committed a crime. Ten years hard labour. Go! Next!" [298]

Korolev was accused of collaborating with an anti-Soviet organization in Germany in order to "subvert a new field of technology." This event may not have been entirely unexpected following the arrest and prison sentencing of Valentin Glushko, another leading Soviet rocket designer three months earlier.[299]

As with Glushko, Korolev received no trial. He was beaten and forced to confess. After receiving a 10-year sentence and having his family property confiscated, he was moved from one prison to another.

In October 1939 he was transferred to the most dreaded of all, the Kolyma forced labor camp in far eastern Siberia made infamous in the West in Aleksandr Solzhenitsyn's publication of

The Gulag Archipelago.

Several thousand prisoners at Kolyma reportedly died from malnutrition, lack of shelter and harsh discipline each month...as many as 30 percent per year. Korolev's five-month-winter ordeal cutting trees, digging and pushing wheelbarrows at a Kolyma gold mine resulted in heart damage, a broken jaw and the loss of all teeth.[300]

Following physically and emotionally grueling Kolyma experiences and numerous failed appeals, Korolev's case was reinvestigated and his sentence was reduced from 10 to 8 years in 1939 when Lavrentiy Beria was replaced by Nikolai Yezhov as Minister of Internal Affairs.

He was then moved to greatly improved conditions at a penal institution known as Central Design Bureau 29 where most occupants were intellectuals, including scientists and engineers, and put to work in charge of wing design for a light bomber.

Korolev was later relocated to another penal institution in Kazan, Siberia, when Germans approached Moscow. Although technically freed in 1942, he voluntarily elected to stay in order to continue work he considered important and he soon became a chief designer for aircraft engines.

Then in 1944, he was moved once again to a penal facility in the Caucasus, where he once again began working as a rocket engineer until his formal release as a prisoner.

Korolev's early death at age 59 might be attributed in part to health problems arising from brutal imprisonment conditions suffered as a victim of an oppressive Stalin regime.

Yet he was never known to speak to anyone about his hard treatment and privations until later, just a few days following a 59th birthday party when late that night after other guests had departed, he confided a sad account to the world's first Earth-orbiting human Yuri Gagarin.[301]

Larry Bell

Wernher von Braun, Germany's Peaceful Space Warrior

By the late 1930s, when some ballistic missile development was occurring at the U.S. Jet Propulsion Laboratory, the Germans were already developing plans for a major Peenemunde Army Research Facility for fearsome V-2 rocket production in a small town located on Usedom Island on their northern coast. Those activities would ultimately be directed by a charismatic and effective engineer...Wernher von Braun.

Wernher had several characteristics in common with his Soviet counterpart, Chief Designer Korolev. As also with American rocketry pioneer Robert Goddard, all began their careers experimenting as rocket amateurs. In addition, although both von Braun and Korolev maintained spaceflight to the Moon and planets as key goals, they received early funding for military missile development, and although terms of punishment were vastly different, both were imprisoned for alleged subversion of those military projects.

Unlike Korolev, who suffered hard labor at the notoriously brutal Siberian camp, following Germany's WWII defeat von Braun was incarcerated by his U.S. captors under incomparably more comfortable conditions in Fort Bliss, Texas, for a mere two weeks.

Following his release along with 126 of his former Peenemunde colleagues, von Braun rose to a level of deserved international fame that Korolev would never know. Accomplishments of the American-German team he led included the development of the Jupiter intermediate-range ballistic missile, the Redstone rocket that launched America's first satellite and first U.S. astronaut Alan Shepard and the Saturn V rocket that enabled 12 fellow Earthlings to walk on the Moon.

Wernher von Braun was born of a noble family on March

23, 1912, in Wirsitz (now Wyrzysk), Poland, which was at that time part of Prussia and the German empire. His rocket interest was kindled by Transylvanian rocket pioneer Hermann Oberth's 1923 *By Rocket into Planetary Space* (English translation). A decade later, while pursuing a mechanical engineering degree, his university VfR Spaceflight Society conducted liquid-fueled rocket motor tests in support of Oberth's work.

Crediting Hungarian engineer Oberth as an important career mentor, von Braun wrote:

> *Hermann Oberth was the first, who when thinking about the possibility of spaceships grabbed a slide-rule and presented mathematically analyzed concepts and designs…I, myself, owe to him not only the guiding-star of my life, but also my first contact with the theoretical and practical aspects of rocketry and space travel. A place of honor should be reserved in the history of science and technology for his groundbreaking contributions in the field of astronautics.*

Von Braun subsequently pursued a doctorate in physics at the University of Berlin, graduating in 1934. That same year his academic group launched two rockets reaching between one and two-mile altitudes. His graduate studies included rocketry research conducted at a solid-fuel rocket station not far from Berlin under the supervision of then-Captain Walter Dornberger, a department head for the German armed forces Ordinance Department.

During this time period, the National German Workers Party (NSDAP, or Nazi party) came into power and moved rocketry into the national agenda. His 1934 thesis titled *Construction, Theoretical, and Experimental Solution to the*

Problem of Liquid Propellant Rocket was kept classified by the German government and not published until 1969.

In the early 1940s, von Braun moved to the new Peenemunde facility as its technical director under the command of Dornberger where his group, in combination with the Luftwaffe, developed liquid-fuel rocket engines for aircraft and jet-assisted takeoffs. Even more significantly, Peenemunde became the development center for a new A-4 ballistic missile which was to become better known as the V-2.

Von Braun was briefly imprisoned on espionage charges for resisting an attempt by Gestapo Chief Heinrich Himmler to take control of the V-2 project. He was reportedly released under Hitler's personal order soon after Germany invaded Poland to start World War II in 1939.[302]

A severe wartime labor shortage in 1943 prompted a plan to use slave labor at the Peenemunde V-2 rocket factory located at Mittlelwork. As with other slave labor operations, brutal treatment of working prisoners produced many tragic casualties. Although von Braun admitted visiting the plant on many occasions and called conditions there repulsive, he claimed never to have witnessed any beatings or deaths directly. However, he admitted that by 1944 it had become clear to him that these incidents had, in fact, occurred.[303]

Adolf Hitler signed an order on December 22, 1942, approving mass production of the V-2 to target London. British and Soviet intelligence agencies soon became aware of the program. Over the nights of August 17 and 18, 1943, the RAF Bomber Command's Operation Hydra dispatched 596 aircraft which dropped 1,800 tons of explosives on the Peenemunde facility. Although it was later salvaged and most of von Braun's team escaped unharmed, the raids killed his engine designer and chief engineer and succeeded in interrupting the program.[304]

Historian Michael Neufeld quotes Von Braun in his book *Wernher von Braun: Dreamer of Space, Engineer of War* expressing unhappiness upon hearing news of the London raids. Representing his interests in rocket applications for space travel rather than war, he reportedly said the rocket worked perfectly, except for landing on the wrong planet.[305]

Following the end of WWII in 1945, Von Braun and his rocketry team (including his brother Magnus) voluntarily surrendered to American forces as part of Operation Paperclip, and he eventually became technical director of the U.S. Army Ordnance Guided Missile Project in Huntsville, Alabama, as well as director of the NASA Marshall Space Flight Center from 1960 to 1970. He also later became vice president of the aviation company Fairchild Industries Inc. and a National Space Institute founder.

In addition to crediting inspirational influences of Hungarian Hermann Oberth, von Braun also acknowledged the importance of lessons taken from technical journals of an American rocketry pioneer named Robert Goddard. Commenting on Goddard designs, he observed:

His rockets...may have been rather crude by present-day standards, but they blazed the trail and incorporated many features used in our most modern rockets and space vehicles.[306]

Robert Goddard, America's Father of Spaceflight

German and Russian liquid-fueled rocket development began years after Robert Goddard launched the world's first one weighing 16 pounds on March 16, 1926, from his Aunt Effie's farm in Auburn, Massachusetts. This was five years before Johannes Winkler launched Germany's first at Dessau in 1931, and seven years before the Soviet Union's GIRD-09 1933 success in the Nakhahino woods.

As Goddard described his launch to sponsor Charles G.

Larry Bell

Abbot at the Smithsonian Institution:

> *After about 20 seconds the rocket rose without*
> *perceptible jar, with no smoke and with no*
> *apparent increase in the rather small flame,*
> *increased rapidly in speed, and after describing a*
> *semicircle, landed 184 feet from the starting*
> *point—the curved path being due to the fact*
> *that the nozzle had burned through unevenly,*
> *and one side was longer than the other. The*
> *average speed, from the time of flight measured*
> *by a stopwatch, was 60 miles per hour. This test*
> *was very significant, as it was the first time a*
> *rocket operated by liquid propellants traveled*
> *under its own power.*[307]

Although each of these early launches entailed extremely tiny and crude devices by today's standards, paraphrasing the immortal words of Neil Armstrong upon reaching the Moon's surface on July 20, 1969, those small steps indeed led to giant leaps.

Some financial support from the Guggenheim family—thanks to an endorsement from American aviation hero Charles Lindberg—enabled Robert Goddard to leave his professorship position at Clark University in Worcester, Massachusetts, and move his rocket development work to Roswell, New Mexico, in 1931. However, lack of success in obtaining U.S. military interest along with bad economic depression conditions which reduced existing sponsorship support forced him to return to academia. Ironically, only 12 days after the first successful GIRD launch, Goddard received a letter from the Acting Navy Secretary stating:

> *Because of the great expense that would be*
> *entailed in development of the rocket principle*

276

> *for ordinance and aircraft propulsion, which under present stringency of funds appears hardly warranted, the Department regrets it is not in a position to further such development.*[308]

Goddard received a similarly discouraging rejection letter seven years later in 1940 from the U.S. Army Air Corps.

A letter from Brigadier General H. Brett stated:

> *While the Air Corps is deeply interested in the research work being carried out by your organization under the auspices of the Guggenheim Foundation, it does not, at this time feel justified in obligating further funds for basic jet propulsion research and experimentation.*[309]

Goddard tended to eschew publicity, sharing many of his most imaginative ideas only with trusted friends and groups. He did, however, publish a March 1920 letter to the Smithsonian which discussed possibilities of photographing the Moon and planets from rocket-powered fly-by probes, sending messages to distant civilizations on inscribed metal plates, the use of solar energy in space, and the idea of high-velocity ion propulsion.

Those early ideas, which were generally regarded as very radical at the time, drew strongly sensationalized media publicity and criticism. A front-page January 12, 1920, New York Times story titled *Believes Rocket Can Reach Moon,* was followed days later with an editorial that scoffed at Goddard's proposals. The article argued, among other disagreements, that:

> *[A]fter the rocket quits our air and really starts on its longer journey, its flight would be neither accelerated nor maintained by the explosion of*

the charges it then might have left. To claim that it would be is to deny a fundamental law of dynamics, and only Dr. Einstein and his chosen dozen, so few and fit, are licensed to do that.

Then, to add more insult to injury, the *New York Times* challenged Goddard's understanding of Newton's fundamental laws. Asserting that thrust can't occur in a vacuum it concluded:

That Professor Goddard, with his "chair" in Clark College and the countenancing of the Smithsonian Institution, does not know the relation of action and reaction, and of the need to have something better than a vacuum against which to react—to say that would be absurd. Of course he only seems to lack the knowledge ladled out daily in high schools.[310]

Although Robert Goddard's rockets never achieved great altitudes, that wasn't really his goal.

Rather, his work concentrated upon perfecting liquid-fueled engines along with reliable and accurate guidance and control subsystems which would eventually achieve high altitudes without tumbling in the thin atmosphere and provide stability for sensitive experiments and other payloads current and future rockets would carry.

The father of American spaceflight was on the verge of developing larger rockets capable of reaching extreme altitudes when World War II intervened to change everything.

And yes, such devices later proved to work very well in the vacuum of space after all.

The Undeclared Space Race

An October 4, 1957, front-page *New York Times* headline in half-inch capital letters carried a story that was being reported all over the world: *SOVIET FIRES EARTH SATELLITE INTO SPACE; IT IS CIRCLING THE GLOBE AT 18,000 MPH; SPHERE TRACKED IN 4 CROSSINGS OVER US.* That orbit repeated more than 1,400 rounds before Sputnik-1 stopped chirping out its ominous presence and burned up in the atmosphere three months later.

Or as an October 7th Manchester Guardian editorial titled *Next Stop Mars* described the event:

> *The achievement is immense. It demands a psychological adjustment on our past towards Soviet society, Soviet military capabilities and— perhaps most of all—to the relationship of the world with what is beyond.*

It went on to more ominously speculate that:

> *The Russians can now build ballistic missiles capable of hitting any chosen target anywhere in the world.*

The concept of launching satellites to orbit wasn't new. Tsiolkovsky's 1903 calculations showed that a device launched at a certain velocity could overcome the pull of Earth's gravity and achieve orbit. Slightly more than a half-century later, Korolev's team had developed a rocket capable of accessing that necessary 8,000 meters per second orbital trajectory.

That goal had been studied and pursued in America as well. During the mid-1940s, the U.S. Army Air Corps asked major airframe companies to submit secret competitive proposals for the design of an Earth-orbiting satellite. This led to funding a newly formed Project RAND (Research and Development) in Santa Monica, California, to study the matter. Their report concluded that:

> *The achievement of a satellite craft by the United States would inflame the imagination of mankind, and would probably produce repercussions in the world comparable to the explosion of the atomic bomb.*

During early 1954, the United States began considering plans to place a small satellite in orbit as its 1957-58 International Geophysical Year (IGY) contribution. In response, Wernher von Braun's team at the Army's Redstone Arsenal in Huntsville began meeting with George Hoover of the Office of Naval Research to accomplish this goal using existing Army Ordnance weapons technology. Their proposed solution, Project Orbiter, was to be an Army-Navy-Air Force design.

The Soviets were concerned about America launching an IGY satellite before they did. Towards the end of 1953, the R-7 rocket made by the Chief Designer's team could launch a 5-ton ICBM warhead and could also easily orbit a 1.5-ton satellite.

But what sort of satellite? The Soviet Academy of Scientists was presented with various options for IGY. One possibility was a living organism such as a dog. Another was to fly around the Moon and photograph the side hidden from Earth. The big priority, however, was to beat the Americans. Too ambitious of a plan would fail that purpose.

They finally settled with that plain polished metal sphere carrying only a radio transmitter, batteries and temperature-measuring instruments. It worked, and as a Pravda headline proclaimed, *World's First Artificial Satellite of Earth Created in the Soviet Union.*

One month later on December 6th, the first American Vanguard Program launch attempt ironically designated TV-3 (for Test Vehicle 3) failed before world television cameras. After rising but a few feet off the ground it ignominiously sagged back, buckled, burst into flame and tossed its tiny three-pound satellite still transmitting a short distance away. *Pravda* reproduced a front-page London *Daily Herald* photo showing the explosion with a superimposed headline which in translation read *"OH, WHAT A FLOPNIK!"*

Far more fortunately for the United States, following a Sputnik-2 launch, a von Braun group from the Army Ballistic Missile Agency developed a Jupiter C launch vehicle which successfully placed a 28-pound Explorer-1 satellite in orbit on January 31, 1958. The scientific benefits proved historic when its onboard instruments first discovered now-famous Van Allen radiation belts.

Technological bragging tables turned on the Russians after a launch failure three months later on April 27, 1958, with its 1.3-ton Sputnik-3 payload aboard. Although a successful follow-up launch was soon achieved, its replacement Sputnik-3 satellite missed an opportunity to further map the Van Allen belts due to a satellite positioning failure.

Following Sputnik, a seesaw series of competitive orbital launch successes and flubs ensued on both sides.

The first three Soviet attempts to place satellites on a Moon failed to reach an Earth-departure orbit. The fourth, a January 2, 1959, Luna-1 launch missed its target by 6,000 kilometers yet succeeded in orbiting the Sun. Luna-2 launched on September 12, 1959, and became the first spacecraft to make contact with the Moon or any other celestial body. Luna-3, launched only three weeks after Luna-2, photographed the far side of the Moon never before seen by humans.

Meanwhile, Americans were realizing some launch misses

and hits as well. These began on August 17, 1958, when a first stage Air Force Thor-Able spacecraft of the subsequent Pioneer Program malfunctioned 77 seconds after launch from Cape Canaveral. Although Pioneer-1 third stage missed the Moon, it set a distance record by traveling some 113,854 kilometers into space. Pioneer-3 then provided important data about the outer Van Allen radiation belt. Notwithstanding these significant achievements, there were seven straight Pioneer Moon misses through 1960.

As for moving targets, Mars and Venus destinations presented far more complicated trajectory guidance challenges than the Moon. Two October 1960 Soviet Mars probe failures to reach Earth's orbit were followed by seven straight failed Venus probes—five by Russia and two by America—between February 1961 and September 1962.

The first failed Soviet Mars launch. which occurred at the time of the Khrushchev-Kennedy standoff over Cuban missile emplacements, might very well have led to a little-publicized but hugely larger competitive Russian-United States disaster. Just as the spacecraft was being prepared for launch, Korolev's team was ordered to immediately remove it and abort the mission so that a military ICBM could use the site in response to a U.S. thermonuclear strike.

Although the issue soon appeared to be settled via diplomatic channels and launch preparations were allowed to proceed, that didn't end the problem.

On October 24, 1962, still in the middle of the crisis, the Mars launcher exploded into so many pieces during ascent that observers at the U.S. Ballistic Early Warning System feared that a Soviet nuclear attack might have commenced. The crisis was averted when computers which assess trajectory and impact points reported a false alarm within seconds.

The international nuclear war scares subsided on October 27, 1962, when Khrushchev announced he would dismantle the missiles in Cuba and return them to the USSR. One week later a Soviet spacecraft designated Mars-1 made the first (unintentional) Red Planet flyby after losing communications. Nevertheless, it accomplished the impressive feat of traveling 106 million kilometers and sending back 61 batches of data until March 21, 1963.

The mutually disclaimed U.S.-Russian competition intensified over the next several years, with each nation anticipating and closely monitoring activities of the other. Both continued to experience significant, if incremental, successes and failures.

A Russian Zond-1 spacecraft launched on April 2, 1964, and reached the vicinity of Venus on July 20, although a radio failure resulted in no returned data. Venera-3 launched on November 16, 1965, and accomplished the first Venus impact on March 1, 1966. Zond-2 launched on November 30, 1964, and demonstrated use of the first electric thrusters for attitude control.

On the American side, although its instruments failed, on April 23, 1962, Ranger-4 became the first U.S. spacecraft to impact the Moon. Launched on October 18 of that year, Ranger-5 missed the Moon by 700 kilometers. Ranger-6 hit the Moon on January 30, 1964, but the TV camera didn't work.

Rangers 7, 8 and 9 returned marvelous pictures covering over 400,000 square kilometers of the lunar surface between 1964 and 1965. And while Mariner-1 failed to reach Venus, Mariner 2 flew within 35,000 kilometers of the planet. Mars-3 missed Mars, but in July 1965, Mars-4 sent back spectacular TV pictures of its cratered surface from a distance of 9,844 kilometers.

The 1970s witnessed more historic interplanetary achievements by both Russia and the U.S. In 1971, five years after Korolev's death, two Soviet capsules released by Mars-2 and Mars-3 crashed into the Martian surface on November 27 and

December 2, respectively. The first of those events occurred just two weeks after the American Mariner-9, developed by NASA's CalTech Jet Propulsion Laboratory, orbited around the planet throughout a dust storm, then sent back detailed pictures of the surface until January 1972.

Although Korolev never lived to witness Russian probes reaching Mars, he did experience a personal triumph which commenced a transformational new era of human space exploration. As the *New York Times* exclaimed on April 12, 1961, once again in bold front-page headlines:

> *SOVIET ORBITS MAN AND RECOVERS HIM; SPACE PIONEER REPORTS: 'I FEEL WELL': SENT MESSAGES WHILE CIRCLING EARTH.*

The newspaper followed up with an editorial the next day prophesying that the "flight will be hailed as one of the great advances in the story of man's age-old quest to tame the forces of nature." Pravda declared it a *GREAT EVENT IN THE HISTORY OF MANKIND.* The Communist Party seized upon the event as a triumph over capitalism.

This was not generally greeted as good news by the majority of Americans, and particularly not by those connected with the U.S. space program. Gagarin's one-hour, 48-minute full-orbit demonstration aboard a Vostok-2 spacecraft on April 12, 1961, eclipsed a 15-minute-long suborbital launch of Navy jet pilot Lt. Col. Alan Shepard on May 5th of that year which reached a 167-mile altitude and traveled 302 miles downrange.

Unfortunately for American history, Shepard's flight, which was originally intended to be a full-orbit launch aboard a Redstone rocket, was delayed multiple times following von Braun's decision that another test flight was needed after a previous one traumatized its passenger...a chimpanzee named

Ham.

George Low, then chief of manned space flight, recalled a conversation between newly appointed NASA Administrator James Webb and his deputy Robert Seamans, who had just testified on the state of their efforts before the House Committee on Science and Astronautics on the day before Gagarin's flight. Webb and Seamans decided not to show a film following Gagarin's historic world spectacle. The movie featured recovery of the dazed chimpanzee Ham two and one-half months earlier.

Low remembered the conversation prudently concluding:

> *...it would not be in our best interest to show how we had flown a monkey on a suborbital flight when the Soviets had orbited Gagarin.*

Although the distinction of being the first American to orbit Earth ultimately went to John Glenn, fortune later beamed more brightly upon Alan Bartlett Shepard Jr. upon becoming the only Mercury astronaut to walk on the Moon on the Apollo 14 mission.

Gus Grissom suffered far worse fortunes. He nearly drowned when explosive bolts fired unexpectedly, blowing the hatch off his Liberty Bell capsule during splashdown following the second suborbital Project Mercury-Redstone flight on July 21, 1961. A catastrophic fire ended the lives of Apollo Astronauts Grissom, Roger Chaffee and Ed White during a January 27, 1967, test.

Russia and America had both been conducting animal tests to determine if humans could survive launch and re-entry stresses in addition to weightless orbital conditions. Soviet space scientists had been experimenting with canines since at least 1951. Dogs Dezik and Tsygan were sent to a 100-kilometer altitude that year using the same pod that carried Laika. The

United States had experimented with monkeys since the 1950s.

Not known to Westerners until mentioned in a 1994 publication, Russian space canine experiments had not always led to successes. A July 28, 1960, Vostok prototype flight carrying dogs Chaika and Lisichka failed when the launch vehicle exploded. More fortunately, an 18-orbit flight the following month ended far better for dogs Belka and Strelka, who became the first creatures to return alive.

A few weeks later, the Communist Party approved a request to launch a human. Not even the catastrophic October 24, 1960, launch pad explosion nor the failure of another December 1st launch failure which killed two other dogs halted plans to go ahead.

As always, top-secret Chief Designer Korolev was kept off to the side away from public fanfare following Gagarin's triumphant return. Worse for America, NASA achievements had been sidelined altogether.

Alternate Destinies, Dead Ends / Limitless Futures

A chastened America got the Gagarin flight message without need of the dazed monkey film.

Revelations of Soviet ballistic missile advancements and geopolitical implications made international headlines in 1962 when intended placements in Cuba of R-16 ICBMs triggered a fearsome Kennedy-Khrushchev confrontation. Their range of about 2,200 kilometers combined with their basing in Cuba and western Russia posed a major threat not only to the United States, but also to bomber bases in Europe and Asia.

Both countries at that time were aware they were being spied upon from overhead by the other. Korolev's design bureau had developed and launched four spy satellites beginning with Kosmos-4 (later known as Zenit) on April 26, 1962. By that time, the United States had already been flying its own spy satellites for nearly two years under a top-secret Corona program that was declassified in 1995.[311]

The Zenit cameras could cover the entire United States in about twenty-five orbits and reportedly determine the number of cars in a parking lot. These were ideal capabilities for ICBM site mapping and monitoring. Corona could presumably do the same.[312, 313]

The combined timing of the Bay of Pigs and Gagarin headlines put great pressure on Kennedy to demonstrate resolute leadership. On May 25, 1961, just slightly less than six weeks following Gagarin's catapult into the Space Hall of Fame, he did so, announcing before a special joint session of Congress that an American astronaut would be safely sent to the Moon before this decade is out.

The American national security implications of both events leading up to Kennedy's bold declaration were made crystal clear in Soviet leader Nikita Khrushchev's November 22, 1957 statement during an interview with publisher Randolph Hearst Jr. three months after his country had conducted a R-7 missile simulated warhead launch:

> *The Soviet Union possesses intercontinental ballistic missiles. It has missiles of different systems for different purposes. All our missiles can be fitted with atomic and hydrogen warheads. Thus, we have proved our superiority in this area.*[314]

Even more pointedly, Khrushchev disparaged the defensive potency of U.S. naval power in a September 7, 1958, letter to President Eisenhower, stating:

> *In the age of nuclear and rocket weapons of unprecedented power and rapid action, these*

once formidable warships are fit, in fact, for
nothing but courtesy visits and gun salutes, and
can serve as targets for the right type of
rockets.[315]

The issue of a purported U.S.-Soviet missile gap had entered the 1960 presidential campaign when candidate Kennedy claimed that the Russians had built up a substantial lead. President Eisenhower, who had access to spy satellite and secret U-2 overflights, knew differently, but refrained from saying so to protect those sources. Early 1960 CIA Russian ICBM estimates put the total at 35 missiles, with the number growing from between 140 and 200 by mid-1961...not the thousands that some were wildly speculating.[316]

Conditions for American dominance in a missile race had grown grim. In response to a reporter's question regarding when the United States might perhaps surpass Russia in this field, President Kennedy providently observed, "the news will be worse before it is better." He was right, On August 6, 1961, Gherman Titov's seventeen-orbit Vostok-2 flight topped Gus Grissom's suborbital Redstone-4 flight of July 21st.

Kennedy's September 12, 1962, commitment to human lunar exploration at Rice University's stadium left no doubt that this was to be a competitive race dedicated to demonstrating U.S. technological supremacy over the Russians. Kennedy warned:

Within these last 19 months at least 45 satellites
have circled the Earth. Some 40 of them were
made in the United States of America and they
were far more sophisticated and supplied far
more knowledge to the people of the world than
those of the Soviet Union.

288

What had previously been a tacit matching of wits and capabilities involving post-war rivals had become an officially recognized race between superpowers. The situation literally began to look up for America when a Mercury-Atlas 6 rocket launch of Friendship-7 carried Colonel John Glenn on three orbits on February 20, 1962.

Three months later, Mercury-Atlas 7 carried Scott Carpenter on three orbits, splashing down 420 kilometers beyond the target area due to reentry errors. On October 3, 1962, Mercury-Atlas-8 Astronaut Walter Schirra did even better, splashing down within 7.24 km of a recovery ship following 6 orbits. That record, in turn, was beaten by Gordon Cooper on May 16, 1963, on Mercury-Atlas 9, which landed 6.4 km from ship following 22 orbits.

Meanwhile, on the other side, Soviet Cosmonaut Andrian Nikolayev had performed 65 orbits on August 11-15, 1962, aboard Vostok-3, while Pavel Popovich, who launched the next day aboard Vostok-4, did 48, the two spacecrafts flying in orbits within 5 kilometers of each other. On June 14-19, Valery Bykovsky, aboard Vostok-5, passed within 5 kilometers of Russia's (and the world's) first woman, Cosmonaut Valentina Tereshkova, aboard Vostok-6. The high international prestige stakes of a race to the Moon competition weren't lost on Khrushchev, Korolev and others in the Soviet Union. As quoted by Korolev associate Oleg Ivanovsky:

> He would tell us that 'the Americans are at our heels, and the Americans are serious people.' He wouldn't use the word 'Amerikantsi' but 'Amerikan-ye' as if these weren't just American residents but the entire American culture we were competing with. He didn't mean this as an insult but as a show of respect for the competition.[317]

Korolev is believed to have made his first serious proposal for a manned Moon mission during an April 6, 1956, speech to the Soviet Academy of Scientists...more than a year prior to the Sputnik-1 launch. He stated:

> *This real task is to fly to the Moon and back from the Moon. This task is most easily solved by starting from Earth. Somewhat more difficult will be returning to Earth that will be on a satellite or rocket that goes to the Moon. But it must not be believed that the proposals I am making are extremely remote.*[318]

Korolev's writings made his primary objective clear:

> *[These] first studies of the Moon and interplanetary space at distances that reach 400-500 thousand kilometers will also create the necessary prerequisites/premises for the penetration of man into interplanetary space, the Moon and the planets.*[319]

Buzz Aldrin told me prior to the release of news reports following Apollo that he believed that the Russians had secretly planned to beat America to the Moon with their astronauts. In 1988 I was among the first Americans to be invited, along with several of my space architecture graduate students, to visit the facility in Moscow and to see a mockup of a lunar surface habitat module they had been planning for exactly this purpose.

The Lunokhod (Moonwalker) program, which was intended from the beginning to support manned surface missions, had produced remote-controlled rovers that predated all others to be deployed on a celestial body. Launched aboard powerful Proton-K rockets, they were controlled from a network of ten ground-based facilities containing Earth satellite vehicle tracking equipment along with command/controls for Soviet near-space civil and military operations.

The first of the total two successful Lunokhod rover deployments (Luna 17) reached the Moon in the Sea of Rains on November 17, 1970. Measuring 4 ft 5 inches high, it carried four television cameras, special extendable devices to collect lunar soil for density and mechanical property tests and a cosmic ray detector. Over its 322 Earth days of operations, it traveled 6.5 miles, returned more than 20,000 TV images and performed a series of 25 X-ray fluorescence soil analyses at 500 different locations.

Lunokhod-2 (Luna 21), a more advanced robot, landed on January 15, 1973. Equipped with three slow-scan TV cameras mounted high on its rover for navigation, it returned images to ground controllers on Earth who sent real-time driving commands. Scientific instruments included a soil mechanics tester, solar x-ray equipment and a French-supplied photodetector for laser detection experiments. It returned about 80,000 pictures over five months of operations.

Major achievements of Russia's Moon program also included three robotic sample return missions: Luna-16 (September 1970), Luna-20 (February 1972) and Luna-24 (August 1976). Altogether they collected and returned 0.326 kg of lunar surface materials.

Space Pathways to Peaceful Cooperation

Although not broadly known, two years before his assassination on November 22, 1963, Kennedy had proposed a possible joint cosmonaut-astronaut lunar mission to Khrushchev at a Vienna luncheon. After some consideration, Khrushchev rejected the offer.

Kennedy proposed the idea again just two months before his

death at a September 20, 1963, United Nations General Assembly appearance, stating:

> *Space offers no problems of sovereignty...Why, therefore, should man's first flight to the Moon be a matter of international competition? Why should the United States and the Soviet Union, in preparing for such expeditions, become involved in immense duplications of research, construction, and expenditure? Surely we should explore whether scientists and astronauts of our two countries—indeed of all the world—cannot work together in the conquest of space, sending some day in this decade to the Moon not the representatives of a single nation, but the representatives of all of our countries.*[320]

In any case, while neither Kennedy nor Khrushchev would live to see their astronauts and cosmonauts joining together on evening Moon strolls, both nations would leave separate tracks. Twelve Apollo explorers beginning with Neil Armstrong and Buzz Aldrin would impress human footprints on its surface, while robotic Soviet explorers would leave mechanical track prints before the end of the Apollo program in 1974.

International pathways converged as the 1970s and 80s ushered in a new era of international space cooperation. With Apollo ended, Russian and U.S. attention shifted from lunar exploration to Earth orbital studies of human adaptation and mitigations. These studies focused on extended weightlessness, influences of weightlessness and space vacuum upon materials and physical/mechanical processes and ways to enhance human safety and performance during future long-term, multi-year missions.

Many nations accepted invitations to join and invest in

common, peaceful enterprises of discovery in orbit and beyond. As a result, the world continues to witness benefits of multicultural engagement rising far above limited boundaries of interest defined on surface maps. This ongoing quest of discovery may ultimately prove to be the greatest and most rewarding space exploration challenge of all.

Global audiences watched on July 17, 1975, as TV images showed Soviet Cosmonauts Alexei Leonov and Valery Kubasov shaking hands with NASA Astronauts Thomas Stafford, Vance Brand and Donald (Deke) Slayton high above the Atlantic Ocean. The historic docking of an American Apollo Command and Service Module (CSM) and Russian Soyuz vehicle mated critical mechanical pressure seals developed separately for the first time. This feat occurred after the two spacecrafts delivering them had made gradual trajectory changes over a two-day period after being launched within hours of each other. One had departed from the Kennedy Space Center in Florida…the other from the Baikonur Cosmodrome in Kazakhstan.

The political timing of the Apollo-Soyuz mission was no accident, highlighting a new policy of détente, a symbolic act of peace, between superpowers. The United States was engaged in a Vietnam ground war which, given Russia's proxy involvement in the conflict, was adding to existing Cold War tensions. The government-controlled Soviet press had been highly critical of America's Apollo program, printing on one occasion that:

> …the armed intrusion of the United States and Saigon puppets into Laos is a shameless trampling underfoot of international law.

Soviet leader Leonid Brezhnev shifted that public position to extol peaceful diplomatic benefits of the Apollo-Soyuz experiment.

He told the world:

Larry Bell

The Soviet and American spacemen will go up into space for the first major joint scientific experiment in the history of mankind. They know that from outer space our planet looks even more beautiful. It is big enough for us to live peacefully on it, but is too small to be threatened by nuclear war.

Apollo-Soyuz, the first joint U.S.-Russian space flight, was to be the last flight for an Apollo spacecraft. It was also to be the last manned U.S. space mission until the first Space Shuttle launch in April 1981, providing important experience for future Shuttle-Mir and International Space Station (ISS) programs that followed.

NASA's Space Shuttle program emerged when the agency convened a task group in 1968 to begin planning beyond Apollo. The priority centered upon developing a reusable Earth-to-orbit transportation system which could support a space station with round-trip crew and cargo delivery.

The program was formally launched by President Nixon's administration on January 5, 1972, with goals of "transforming the space frontier...into familiar territory, easily accessible for human endeavor." North American Rockwell (later Rockwell International, now Boeing), the same company responsible for building the Apollo Command/Service Module, was awarded the development contract.

Over the course of its operations between 1981 and 2011, the five-Shuttle-fleet program accomplished 135 missions. The longest, STS-80, lasted 17 days, 15 hours, and the final flight, STS-135, occurred on July 8, 2011.

Two of the Shuttle orbiter missions suffered catastrophic disasters which killed a total of 14 crew members. The STS-51-L Challenger launch, which exploded 73 seconds after liftoff on January 28, 1986, was caused by the failure of a connecting seal on a long solid-fuel rocket. STS-107 Columbia was lost approximately 16 minutes before its expected landing on February 1, 2003, when a piece of hard insulating foam shed from its large external tank during launch punctured a reentry heat protection tile on its wing edge.

The Soyuz modules originally developed to carry cosmonauts to Salyut space stations and Mir are currently used for ISS. A minimum of two are docked with the ISS at all times to provide assured contingency departure for a crew of six (three passengers each).

The ISS core module providing its primary crew life support systems is based upon the engineering developed for the world's first space station, Salyut-1, which was launched in 1971. Salyut 1, in turn, was originally developed by Korolev's design bureau undercover for a military space station secretly known as Almaz.

Cosmonauts from the Communist bloc and non-American astronauts from the West later repeatedly set time-in-orbit endurance records aboard Salyut 4, 6 and 7 from April 1982 up through early 1991.

Skylab, America's first space station, orbited Earth from 1973 to 1979 and was created using a converted third stage of a Saturn V Moon rocket outfitted with two decks as a habitat and orbital workshop. Strictly a NASA operation, the facility was spacious even by current standards and provided a large solar observatory and experiment area. A Crew Service Module (CSM) converted from the second stage, a smaller Saturn 1B booster, provided crew transport and emergency means to rapidly return to Earth.

Skylab's three total three-crew missions logged 513 man-days in orbit and accomplished thousands of experiments covering many different disciplines. Its orbit was allowed to slowly decay causing the facility to burn up on reentry five years after the last crew had returned home.

Larry Bell

The Mir space station (which translated, combines words like world, peace and village) ultimately served as a true symbol of cooperation between the people of Russia and the United States following a half-century of mutual antagonism. Over its 15 years of operation from the time its first module was launched on February 20, 1986, Mir hosted 125 cosmonauts and astronauts from 12 different nations who conducted approximately 23,000 scientific experiments.

America's Space Shuttles docked with Mir seven times. While prior to the Shuttle-Mir missions, experiments and supplies were provided exclusively by Russian "Progress" cargo vehicles, several Shuttle-Mir missions supported commercial services.

As the world's first modular space station, Mir's cluttered outside appearance has been variously characterized as a dragonfly with wings outstretched, a prickly hedgehog and a 100-ton Tinker Toy. Nevertheless, Mir served as a beautiful symbol of international cooperation in space science and collegiality over a course of 86,000 total orbits.

Ironically, on the very same day Mir's life ended as fragments in a watery grave on March 23, 2001, Russia expelled four U.S. diplomats and threatened to expel 46 more in retaliation for America's expulsion of 50 of theirs for suspected espionage.

Mir's demise un-coincidentally coincided with the beginning of the most complex international scientific and engineering project in history and the largest human structure ever to be put in space.

Planned and operated by five different agency partners representing 15 different countries, the International Space Station serves a variety of purposes: a laboratory for biological, material and other sciences; an observation platform for astronomical, environmental and geological research; and a stepping-stone towards future space exploration. Responsibilities and investments are divided among NASA, Russia's Federal Space Agency Roscosmos, the European Space Agency, the Canadian Space Agency and the Japan Aerospace Exploration Agency.

The scale of ISS is immense, spanning the area of a U.S. football field including end zones. Weighing a total of nearly a million pounds and orbiting at five miles per second the complex provides more livable space than a conventional five-bedroom house where a six-person expedition crew typically remains onboard from between four to six months. For comparison, ISS is nearly four times larger than Mir, and about five times larger than Skylab.

During a visitation with a docked Space Shuttle, the combined ISS complex has supported a total of 13 people for several days. This crew size was temporarily reduced to two-person teams after the tragic Columbia Shuttle disaster during which time the crew and supplies could only reach ISS using Russian Soyuz and Progress spacecraft. ISS hosted its first one-year crew in 2015-2016 involving NASA's Scott Kelly and Roscosmos's Russia's Mikhail Kornienko.

While ISS was originally planned to be decommissioned in 2020, international partners have yet to decide whether to grant a NASA request to postpone that schedule to 2024. A key obstacle to NASA's proposal centers upon deteriorated relations following Russian military activities in Ukraine. Just as the ISS demonstrates the substantial advantages that may be gained through international collaboration in space, this circumstance also demonstrates the serious risks attached to all large-scale international space programs.

U.S. dependence on Russian Soyuz vehicles to transport its astronauts to and from the ISS will end as new commercial crew launch and landing services are coming online.

By 2012 the ISS was already being supported by commercial cargo delivery services provided by SpaceX's reusable Dragon spacecraft, followed by Orbital Science's Cygnus spacecraft in late 2013. Subject to safety certification, each company will be able to carry up to 7 passengers.

Founded in 2002 by PayPal co-founder Elon Musk, SpaceX (the "X" referring to Exploration Technology), has already logged several historic achievements: In 2008, SpaceX became the first privately funded company to launch a (Falcon 1) rocket into orbit; the first to successfully orbit and recover a spacecraft (2010); the first to send a spacecraft to the ISS (2012); the first to launch a satellite into LEO (2013); and the first organization, private or government, to successfully return a first stage back to the launch site and accomplish a vertical landing with a rocket on an orbital trajectory (2015).

SpaceX also conducted its first satellite delivery to geosynchronous orbit (NASA's Deep Space Climate Observatory) in 2015. In June 2018, a two-stage, 180-feet-tall SpaceX Falcon 9 booster conducted the first commercial cargo delivery flight to the ISS, and in March 2019, the company successfully launched and docked an unmanned Crew Dragon capsule to the ISS.

SpaceX is developing a larger reusable replacement of its Falcon 9 named the Big Falcon Rocket (BFR) with a lift capacity to Earth orbit of more than 150 tons. Also referred to by the company as an Interplanetary Transport System (ITS), its first test flight has been tentatively scheduled for 2020.

Founded in 1982, Orbital Sciences (Orbital ATK), which developed the Antares Rocket and Cygnus Spacecraft, helped to create the Orion Launch Abort System that would have helped NASA astronauts escape in the event of an emergency involving the now-canceled Ares 1 rocket. Orbital ATK is currently competing with SpaceX for ISS cargo resupply services using expendable foreign-supplied rockets and delivery capsules.

SpaceX was not the first commercial company to develop and demonstrate a reusable rocket booster rather than sacrifice it after a single flight. On January 23, 2016, Blue Origin, a company created by Amazon founder/CEO Jeff Bezos in 2004, accomplished this feat. The re-used launcher had been recovered from an earlier successful launch and landing that occurred only two months earlier on November 24, 2015, at Blue Origin's West Texas test facility.

Blue Origin's vertical take-off and landing rocket design is powered by the company's own BE-3 engine. The company has developed a sub-orbital six-person New Shepard tourist capsule named in honor of Alan Shepard, who piloted America's first suborbital mission. In addition, Blue Origin is also in the process of creating a two-stage heavy-lift orbital vehicle named the New Glenn in honor of John Glenn. The first stage will be powered by seven of Blue Origin's new BE-4 engines.

NASA is also commissioning a Space Launch System (SLS) and an Orion crew spacecraft for missions beyond Earth Orbit. The SLS is described as a super-heavy expendable launch vehicle which is part of NASA's deep space exploration plans, which will eventually include a crewed mission to Mars. It is being built by Boeing, the United Launch Alliance, Northrop Grumman and Aerojet Rocketdyne.

The development and operations of SLS provide for three Block stages. A Block 1 and Block 1B will use four five-segment solid rocket boosters based upon the same design as the four-segment rockets used on the Space Shuttle, but Block 1B will feature a more powerful Exploration Upper Stage (EUS) second stage capable of lifting more than 100 tons to Earth orbit. Block 2 will combine the EUS with upgraded boosters and be able to launch 130 tons to orbit. This capacity is similar to that of the Saturn V which was used on Apollo missions.

The Orion Multi-Purpose Crew Vehicle (Orion MPCV) is a joint NASA-European Space Agency venture designed to carry a crew of four astronauts to destinations beyond low-Earth orbit in connection with such missions as to the Moon, asteroids and Mars. The overall vehicle consists two main modules: a command module which is primarily being built by Lockheed Martin and a service module containing many of the mechanical and life support systems which are being developed by Airbus Defense and Space.

A New International Frontier

Several nations, including some which are relatively new to space era activities, are embarking upon ambitious plans to explore this new frontier of scientific, technological and economic opportunities both in Earth's orbit and beyond.

In November 2013, India launched a Mars orbiter named Mangalyaan to map potential sources of methane plumes which might indicate the presence of a microbe biosphere deep beneath the Martian surface. When asked why the country would invest in such costly programs, Nisha Agrawal, CEO of Oxfam—a confederation of charitable organizations focused on alleviating poverty—told BBC:

> *India is home to poor people but it's also an emerging economy, it's a middle-income country, it's a member of the G20. What is hard for people to get their head around is that we are home to poverty but also a global power...We are not really one country but two in one. And we need to do both things: contribute to global knowledge as well as take care of poor people at home.*

K. Radhakrishnan, chair of the Indian Space Research Organization (ISRO) elaborated:

> *Why India has to be in the space program is a question that has been asked over the last 50 years. The answer then, and now and in the*

future will be: 'It is for finding solutions to the problems of man and society.'[321]

President Xi Jinping has made it very clear he intends to have China establish itself as a space superpower.

China sent its first astronaut into space in 2003, the third country after Russia and the United States to achieve independent manned space travel. In 2013, three Chinese astronauts spent 15 days in orbit and docked with an experimental laboratory as part of Beijing's plan to establish an operational space station by 2022.

The first of three 20-ton modules are scheduled for launch to the station in 2020. The core module, called Tianhe (Harmony of Heaven), will house three astronauts and carry their supplies for a stay of several months. Over the following two years, two laboratories for scientific equipment will be added.

As Chinese Ambassador to the United Nations Shi Zhongjun explained:

> *The Chinese Space Station belongs not only to China, but also to the world…Guided by the idea of a shared future, the [Chinese Space Station] will become a common home for all humankind. It will be a home that is inclusive and open to cooperation with all countries…*

36 teams from around the world have already applied to send experiments to the station.

China is opening up its manned and lunar space programs to international participation with the focus on nations that have not had access to space technology in the past. Originally centered on cooperation with neighbors in Asia, the infrastructure, technology and cultural exchanges with more than 100 nations now span the world from Asia to Europe and Africa.

Larry Bell

The Chinese government has made it clear that they are very committed to human lunar and planetary exploration. In 2013 they launched a robotic rover called Yulu (Jade Rabbit) that surveyed for useful natural resources in the northwest corner of the giant Imbrium Basin, the left eye of the Man in the Moon. Zhao Xiaojin, director of aerospace for the China Aerospace Science and Technology Corporation, described the rover as "a high altitude patrolman carrying the dreams of Asia."

The next stage to follow will likely land a lunar probe, release a Moon rover and return a probe to Earth.

In December 2018, China became the first country to land a spacecraft on the far side of the Moon. In 2020, the Chinese plan to return samples of lunar soil to Earth for the study of lunar chemistry, minerals and precious resources. One of special interest is helium-3 deposited on the lunar surface by the solar wind which might serve as a valuable fuel for future fusion power. If world fusion technology proves successful, it is estimated that each ton of the material would yield energy equal to approximately 50 million barrels of crude oil.

Space affords a natural supply depot stocked with a vast assortment and quantity of potentially useful, even critical, materials that can expand human experience and enterprise. Of these rich resource caches, the Moon is the closest, representing a relatively near-term source and laboratory for rocket propellant, oxygen and water production and an operational base for development and demonstration of other extraterrestrial technologies.

Recent discoveries of large quantities of water on the Moon have excited great interest for a multitude of applications which can greatly reduce dependence upon costly transportation of Earth-delivered consumables. The precious molecule is vital to life for drinking and constitutes a source of oxygen for breathing. It's also a source of hydrogen and oxygen for rocket propellant to fuel ascent vehicles from the lunar surface and transportation to cislunar (Moon-Earth-orbit) space and beyond. It can also provide highly mass-efficient solar and nuclear radiation shielding material for astronauts.

Recent findings obtained from NASA's Mars Reconnaissance Orbiter (MRO) offer evidence that water also exists on Mars, and that some even intermittently flows on its surface. MRO's imaging spectrometer detected darkish streaks up to a few hundred meters in length along with spectral signatures of hydrated minerals which darken and extend down deep slopes during warmer seasons (above minus 10 degrees Fahrenheit).

Since these features fade and eventually disappear altogether in cooler seasons, it is theorized that hydrated salts lower the freezing point of the liquid brine, just as salt causes ice and snow on roads here on Earth to melt more rapidly.

However, it's presently unclear whether the dark streaks are signatures of the salts, or rather, appear as a result of the existence of briny water that periodically wicks up from subsurface sources.

Mars also has a great diversity of other resources, some of which are present and most likely more accessible than on the Moon. For example, carbon dioxide, nitrogen and hydrogen exist on the Moon, but unlike Mars, only in tiny parts per million quantities.

Mars isn't known to have helium-3, but does have an abundance of deuterium, a heavy isotope of hydrogen that could be used for future nuclear fusion. And while oxygen is abundant in lunar soil, it is tightly-bound in oxides such as silicon dioxide (SIO_2), magnesium oxide (MgO), ferrous oxide (Fe_2O_3) and aluminum oxide (Al_2O_3) which require high energy processes to release.

Extended distance and time operations on human space voyages and lunar/planetary surfaces will require that means are afforded to harvest and use extraterrestrial resources for propulsion, life support and perhaps eventually, for construction

Larry Bell

and even commercial export.

Yet if the latter of these possibilities seem truly remote, consider how leading "authorities" viewed notions of heavier-than-air flying machines prior to the Wright Brothers' first flight in 1903 and Robert Goddard's predictions less than two decades later that rockets could reach the Moon, much less that humans would walk on its surface little more than a half-century later.

304

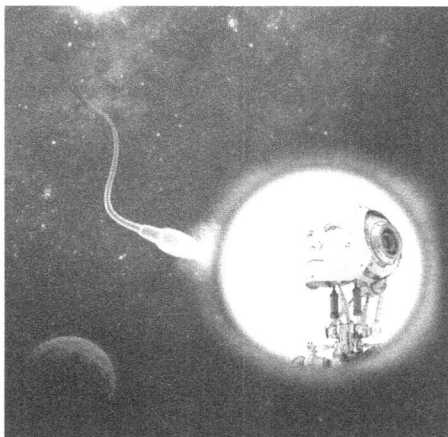

Part Twelve: The Information Revolution

ARTIFICIAL INTELLIGENCE-DRIVEN computational
algorithms and automation in combination with the Internet have
come to be collectively and appropriately referred to as a new
information revolution.

So, what is Artificial Intelligence (AI)? Can humans invent
machines that can really think? Can they learn to compete with
our Sapiens minds' cognitive capacity for general intelligence?

At present, the jury is still out in answering those questions.

According to Andrew Ng, former head of Google Brain and
Baidu Inc.'s AI division, a goal is to enable a computer to do any
mental task the average human can accomplish in a second or
less.

Gary Marcus, former head of Uber Technologies Inc.'s AI
division and currently a New York University professor, believes
that getting to general intelligence—which requires the ability to
reason, learn on one's own and build mental models of the
world—will ultimately take more than AI can achieve.[322]

Thomas Dietterich, former president of the Association for
the Advancement of Artificial Intelligence, views the big AI

machine learning challenge as "to see how far we can get computer systems to learn from data and experience, as opposed to building it in by hand." He observes:

> *The problem isn't that innate knowledge in AI is bad, humans are bad at knowing what kind of innate knowledge to program into them in the first place.*[323]

In any case, AI isn't really a single invention, but rather represents a human tide of innovative progress in many integrative information management and system technology fields. Broadly used, the term is primarily shorthand for any task that computer software can perform just as well, if not better, than humans.

One frequent aspect of AI, termed machine learning, can enable multi-level probabilistic analyses which allow computers to simulate—and even expand—the way the human mind processes data. Here, the process relies upon digital neural networks roughly analogous to those modeled from the human brain.

An AI variant known as deep learning enables computer software to recognize characteristically differentiated information patterns in distinct layers. Here, each neural-network layer operates both independently and in concert with others.

The "deep" in deep learning refers to the number of layers of artificial neurons in a network of neurons. As in biological nervous systems, artificial copies with more layers of neurons are capable of more sophisticated kinds of learning.

Deep learning technology has exploded in popularity since the approach was first described in a 2012 landmark paper posted by researchers at the University of Toronto. An example application is in diagnostic cancer screening which analyzes separate aspects of cell color, size and shape before integrating outcomes. This important advancement of diagnostic medicine can expedite early malignant-stage discovery and treatment.[324]

A Very Brief Early AI History

A group of scientists, mathematicians and engineers from many organizations held the first meeting of a series at Dartmouth College in 1956 to consider a rather preposterous idea. They wondered if it might be possible to create thinking machines that could duplicate or surpass human intellectual capacities.

Two years later, Jack St. Clair Kilby, an electrical engineer at Texas Instruments, came up with another radical idea. He stayed in the lab during a company vacation break and cobbled together a calculating device that combined a transistor, a capacitor and three resistors on a single piece of germanium—the first integrated circuit. It wasn't particularly small—about a half-inch long but narrow—and not very elegant either. With wires sticking out, it resembled an upside-down cockroach glued to a glass slide.

In January 1959, Bob Noyce at Fairchild Semiconductor in Palo Alto, California, developed a precision photographic printing technique using glass as insulation to deposit tiny aluminum wires above silicon transistors without the messy cockroach legs. Kilby's integrated circuit was transformed into a rapidly producible integrated chip with wires 20 times smaller.

The following year, Texas Instruments introduced a Type 502 Flip Flop with one bit of memory that it sold for $450. Weeks later, Fairchild produced its own model which was used by other computer companies, the U.S. Air Force and NASA's Apollo rockets.

One bit soon grew to four, then to 16, then to 64. This capacity increase occurred as the chips continued to rapidly shrink in size. According to a 1965 prediction by Fairchild research director Gordon Moore, now referred to as Moore's Law, the chips' information density would double every 18 to 24 months.[325] By 1969, the TI 3101 64-bit memory chip was priced at $1 a bit. Your iPhone probably has a trillion bits priced at merely pico cents each.

Moore noted that the number of transistors per silicon chip had doubled every year and predicted that the growth would continue over the following decade. By his estimation at that time, microcircuits of 1975 would contain an astounding 65,000 components per chip.

By 1975, as the rate of growth began to slow, Moore revised his time frame to two years. This time his revised law was a bit pessimistic; over roughly 50 years from 1961, the number of transistors had doubled approximately every 18 months. Nevertheless, Moore's estimates were broadly accorded great importance as virtual law.

Moore had based his prediction upon a dramatic explosion in circuit complexity made possible by steadily shrinking sizes of transistors over the decades. Measured in millimeters in the late 1940s, the dimensions of a typical transistor in the early 2010s were more commonly expressed in tens of nanometers (a nanometer being one-billionth of a meter)—a reduction factor of over 100,000.

Transistor features measuring less than a micron (a micrometer, or one-millionth of a meter) were attained during the 1980s, when Dynamic Random-Access Memory (DRAM) chips began offering megabyte storage capacities.

At the dawn of the 21st century, these features approached 0.1 micron across, which allowed the manufacture of gigabyte memory chips and microprocessors that operate at gigahertz frequencies. Moore's law continued into the second decade of the 21st century with the introduction of three-dimensional transistors that were tens of nanometers in size.

Problems with Moore's original capacity estimate which

surfaced in 1975 had encountered a technical snag due to limitations posed by the photolithography process used to transfer the chip patterns to the silicon wafers which used light with a 193-nanometer wavelength to create chips which feature just 14 nanometers.

Although the oversized light wavelength was not an insurmountable problem, it added extra complexity and cost to the manufacturing process. And while it has long been hoped that extreme UV, with a 13.5nm wavelength, will ease this constraint, production-ready EUV technology has proven difficult to engineer.

A roadmap around chip limitations sometimes described as More than Moore applies highly integrated chips which combine a diverse array of sensors and low-power processors. With the growth of smartphones and the Internet of Things, these processors include RAM, power regulation and analog capabilities essential for GPS, cellular, Wi-Fi radios and even advanced microelectromechanical components such as gyroscopes and accelerometers.

Computer chip technology advancements are applying new materials such as indium antimonide, indium gallium arsenide and carbon (both in nanotube and graphene forms) which promise higher switching speeds at much lower power than silicon. Coming soon, expect monolithic 3-D chips, where a single piece of silicon has multiple layers of components built upon a single die.

Prior to the 1970s and 1980s, the only contact most people had with computers was through utility bills, banks and payroll services, or computer-generated junk mail. This condition rapidly changed as integrated circuits on microchips followed by "chip-on-a-chip" advancements dramatically reduced the size and costs over their mainframe predecessors.

Arithmetic, logic and control functions that had previously occupied several costly circuit boards became available in a single integrated circuit which made high-volume manufacture possible. Concurrently, solid state memory improvements eliminated bulky, costly and power-hungry magnetic core systems used in previous generations.

Early advancements were generally created by independent entities and led to the availability of cheap, fast computing, affordable disk storage. Within only a decade, computers became common consumer goods for word processing and gaming. Computing and information storage were contained in personal standalone units.

Personal computers increasingly earned their place in private homes and businesses by the late 1980s. Families, for example, found kitchen computers convenient for storing easily retrievable disk-based recipe catalogs, medical databases for child care, financial records and encyclopedias for school work.

Although predicted to be commonplace before the end of the decade, computers still weren't powerful enough to match more optimistic visions. Due to limited memory capacities, they could not yet multitask, floppy disk-based storage was inadequate both in capacity and speed for multimedia and display graphics were blocky and blurry with jagged text.

Networking technologies created in university computer sciences departments soon led to substantial collaborative software improvements. The resulting emergence of an open-source information culture then spread throughout wide user communities which took advantage of—and also contributed to—common operating systems, programming languages and tools.

It took another decade for computers to mature sufficiently for graphical user interfaces to serve broad, non-technical user markets which gave rise to the Internet. Equipment and user costs dropped dramatically as data catalogs became maintained online and accessed over the World Wide Web rather than stored on floppy disks or CD ROM.

The global digital traffic infrastructure Internet formed as networks became increasingly more uniform and interlinked. Simultaneous increases in computing power and falling data storage costs rapidly expanded worldwide service markets. The Internet, which was popularized for personal email chat and business forums, also became a growing exchange mechanism for computer data and codes.

History has demonstrated that unexpected inventions can rapidly transport our brains, bodies and human potentials along new, uncharted and transformative pathways. As Henry Ford purportedly said: "If I had asked my customers what they wanted, they would have said a faster horse."

Cloud computing is more than just a replacement for faster horses. It is a technological stampede of AI applications which are multiplying and accelerating at warp speed.

As Joe Baguley, the chief of technology officer for VMware EMEA, observes:

> *Just as email rendered the memo obsolete, cloud computing is set to impact the way we do business, offering a competitive advantage to those organizations bold enough to think outside the accepted.*

Baguley adds:

> *The technology is no longer 'nice-to-have,' but is a critical part of any organization's infrastructure. We're seeing more and more businesses relying on mobile devices as the era of 'office working' slowly draws to a close.*[326]

Cloud is not a new idea...it goes back about six decades. In the early sixties there existed the image of delivering computing resources through a global network. J.C.R. Licklider pioneered the idea of the intergalactic computer network in 1969. His aim was to develop an interconnected worldwide network which would enable access to data programs wherever and whenever needed.

Cloud computing rapidly evolved in several directions during the 1970s with the advent of Web 2.0., IBM's breakthrough VM (Virtual Machine) operating system which allowed multiple virtual systems on one physical device.

Fast forward to the 1990s, when Professor Ramnath Chellappa of Emory University and the University of South Carolina defined cloud as the new "computing paradigm where the boundaries of computing will be determined by economic rationale rather than by technical limits alone."

In 2009, computing companies led by Google began to provide browser-based enterprise applications, such as Google App.[327]

While not a new idea, true cloud capabilities are only now beginning to be realized. Revolutionary applications are endless and include management of the digital infrastructure of tomorrow's cities, management and monitoring of driverless car and drone taxi networks and ensuring more efficient operations of farms and power plants.

Cloud will continue to support and empower emerging technologies such as AI and help them to adapt to new platforms such as mobile smartphones, which overtook sales of PCs in 2011. These devices capture lots of unstructured data such as emails, text messages and photos which require more data analysis, time and processing power than most smartphones have.

The same cloud dependency will apply to driverless cars and trucks. Such vehicles will be provided with sensors and cameras that generate huge amounts of data which must be processed in real-time...some within the vehicle, but also by the cloud.[328]

The marvelous confluence of networking, capacity and

storage began in the 1990s. In combination with an open-source culture of sharing that both leads and draws from the Internet, it still remains in the infancy of a yet unknown evolutionary creature.

It is highly speculative—bordering on pure fantasy—to imagine what forms will emerge years, much less decades, in the future. Nevertheless, it's very safe to predict that Moore's innovative lawbreakers will continue to produce computing processors which are more versatile, faster, smaller and energy-efficient in response to ever-growing demands for smarter systems that support and compete with human enterprises.

Artificial Intelligence Race versus the Human Race

The ever-escalating pace of advancements in AI-driven machine learning and automated equipment have placed human societies in competitive marathons with their creators. Computers and robots can already not only perform a range of routine physical activities better and cheaper than humans, but are also increasingly capable of surpassing certain of our cognitive capabilities requiring tacit judgment.[329]

In many respects, the general concept of thinking machines, which was first hypothesized during that 1956 meeting of scientists, mathematicians and engineers at Dartmouth College, is no longer theoretical. Those machines are already beginning to outthink some top human experts in certain very complicated mental challenges.

- By 1997, a "Deep Blue" IBM computer defeated the reigning world chess champion, Garry Kasparov.

- In 2011, Watson, another IBM computer, beat all humans in the quiz show *Jeopardy*.

- In 2016, an AlphaGo algorithm developed by DeepMind, a London AI company, dispatched Lee Sedol, a top player in the ancient and complex board game Go. The algorithm was originally trained on 160,000 games from a database of previously-played games. The program was later upgraded to AlphaGo Zero, which taught itself by playing four million games against itself entirely by trial and error. AlphaGo Zero subsequently annihilated its parent, AlphaGo, 100 games to zero. What it learned in less than one month would have required a decade or two of training for a human.

- In 2017, Libratus software developed at Carnegie Mellon University beat four top players over a 20-day tournament of No-Limit Texas Hold'em poker. The code doesn't need to bluff...it just out-thinks humans.[330]

A technique called Generative Adversarial Networks (GANs) essentially mimics the way we learn from trial and error using positive and negative reinforcement methods. GANs trains competing AI algorithms to challenge each other, learn from mistakes and even to fool one another with convincing deceptions. Both algorithms are trained over a large number of iterations, with each iteration improving the "skill" ability of each.

We have witnessed a worldwide impact of artificial intelligence over just the last few years to the point that it dominates nearly all businesses, investments and even ethical narratives. As a result, two opposing attitudes appear to have emerged: one believing that AI will beneficially augment humans; the other that it will diminish them. Most likely prospects hold that both predictions are true.

There is virtually no likelihood that the AI revolutionary march of encroachment upon the human domain of activities will lose momentum. As Professor Justin Zobel, head of the Department of Computing & Information Systems at the University of Melbourne, Australia observes:

How Everything Happened, Including Us

It is a truism that computing continues to change our world. It shapes how objects are designed, what information we receive, how and where we work, and who we meet and do business with. And computing changes our understanding of the world around us and the universe beyond.[331, 332]

For one, while machines are better at repetitive tasks, people have an advantage when it comes to working with their hands.

McCaney points out that although robotics developers have made a lot of significant advances, getting one to open a door is still a big deal. We also shouldn't expect robots to replace humans in performing a lot of maintenance, plumbing, electrical work or other hands-on jobs in the near future. At best, they will only assist.[333]

Currently, most of us are better than machines at interactions with fellow humans which involve empathy. Sales and counseling in any realm are examples. While an AI assistant can answer factual questions, offering good advice or purposeful listening is a different matter.

Despite advances in natural language processing that enable AI systems to sound human when communicating, the thought behind those words is lacking. This is apparent when it comes to creative forms of communication.

Kevin McCaney emphasizes that for the achievement of general AI—for a machine to be able to think and act like a human—it must be capable of being comprehensively programmed with empathy and creative inspiration essential to create music, to write poetry and fiction and to formulate mathematical proofs.

McCaney cites, for example, an unsuccessful attempt of a London computer scientist who trained an AI bot to churn out poetry in different styles on different topics, based upon seven million words found in 20th century English poetry. Whereas the results sounded like poetry, they lacked subtext or new ideas:

> The frozen waters that are dead are now
> black as the rain to freeze a boundless sky
> and frozen ode of our terrors with the grisly lady shall be free to cry.

As for reading comprehension, Microsoft and Chinese e-commerce giant Alibaba separately reported in 2018 that their AI models had scored slightly higher than humans in a respected Stanford University machine reading test. Scores reflecting percentages of correct answers to 100,000 questions drawn from Wikipedia entries were: Microsoft, 82,650; Alibaba, 82.44; and humans, 82.304.

Nevertheless, Percy Liang, one of the Stanford computer scientists that compiled the test, reported:

> *Even elementary school reading comprehensions are harder, because they often include questions like "Why did X do this?"...So they're a lot more interpretive. We're not even tackling those more open-ended types of questions.*

And while AI is superior at crunching large data sets and recognizing patterns such as the kind of tasks reviewing legal documents could involve, human lawyers and judges remain to be much better at critical thinking and applying lessons learned.[334]

Journalist Alisa Valudes Whyte rhetorically asks in a June 25, 2017, Huffington Post article: *Will AI Best All Humans Tasks by 2060? Experts Say Not So Fast.*[335]

Granted, Whyte acknowledges that computers are learning to compete very rapidly. She notes that by 1997 they were better than humans at chess. Ten years later they were better at driving cars than the average teenager, and they are now better at playing at the Chinese game Go, rated 300 times harder than chess.

Whyte quotes Jay Klein, chief technology officer at Voyager Labs:

> *Every time there's an advancement, we are using that as a ladder to our next step of evolution as human beings. If you look back at the Industrial Revolution, the world didn't stop after some inventions were being made. On the contrary, we continued to evolve.*

CEO and co-founder of Databricks, Ali Ghodsi, doesn't foresee a completely automated future with the human out of the loop any time soon:

> *The way AI is being built is simply not at all in any way the way true human intelligence works, but it can augment us [in some domains] and do a much better job. But humans will still be super critical by 2060.*[336]

Ghodsi recognizes that AI has already gotten extremely good at pattern prediction, enabling so much of the big data revolution that's going on in the industry right now. This, in turn, fuels many other areas where the technology is currently outpacing humans, like image and video recognition. And it's making great gains in areas such as natural language recognition.

AI prowess boils down to excelling at solving defined problems in a highly defined structure, but outside of this structure is where AI will continue to need human influence.

Ghodsi explains:

> *AlphaGo is now beating the human Go players in Go, and that's awesome. But ask the computer to reflect on its victory, and it has no clue what that means. If I asked a human that question, they would have an answer, and the computer would be clueless.*

Ali Ghodsi maintains that humans will continue to be masters of tasks that involve creativity, emotional intelligence or formulating a problem in the first place, instead of just solving one. Computers, on the other hand, require outside programming where there have been few advances.

He observes:

> *If it's an open-ended problem that doesn't have a clear, well-defined structure, then [AI] won't be able to do it.*

AI Influences upon Creative Destruction:

Austrian economist Joseph Schumpeter coined the term "creative destruction" to characterize the way technological progress in the late 1940's improved the lives of many, but inevitably, only at the expense of a smaller few. As improvements to manufacturing processes such as assembly lines benefited the general economy and overall individual lifestyles, craft and artisan producers were displaced.

Others more optimistically argue that while some industries and work roles will indeed fall as casualties of new technologies, they will be replaced by even greater, more open-ended opportunities.

Matthew Randall, the executive director of York College's Center for Professional Excellence, writes in TechCrunch.com that the trend of industrial robots replacing human manufacturing jobs is ultimately a good thing:

> In the last century, we moved from people manually building cars to robots assembling cars. As a result, manufacturers both produce more cars and employ more people per car than before. Instead of performing dangerous tasks, those workers now program the robots to do the dirty work for them—and get paid more for doing so. As long as we've had technology, we've had Luddites who literally destroy technological advancements—and yet, here we are, more productive, with higher quality of living than ever.

Randall argues:

> [In] reality, [robots] will enable us to keep more (and better) jobs at home, to grow our local industry, to improve our lives at the micro and macro levels. With greater automation, efficiency, safety and productivity, the North American manufacturing sector will not only survive, it will showcase the power of our innovation and ingenuity.

He concludes:

> So, will a robot take your job? Maybe, but in return, you—and your children and grandchildren—will likely find more meaningful work, for better pay. Sounds like a good trade-off to me.[337]

Researchers at McKinsey & Company, a leading business consulting firm, project a large need in the future for human cognitive abilities. They conclude in a 2017 report titled *Jobs lost, jobs gained: What the future of work will mean for jobs, skills and wages*:

> *Workers of the future will spend more time on activities that machines are less capable of, such as managing people, applying expertise, and communicating with others. They will spend less time on predictable physical activities and on collecting and processing data, where machines already exceed human performance. The skills and capabilities required will also shift, requiring more social and emotional skills and more advanced cognitive capabilities, such as logical reasoning and creativity.*[338]

The McKinsey & Company researchers determined that as greater percentages of populations live longer, significantly larger new demands will result in a range of healthcare occupations, including doctors, nurses, health technicians, nursing assistants and personal home-care aids.

New careers and jobs will also be created in technology development and information technology services. While this will be a relatively small number of jobs compared to employment in healthcare and construction, the jobs will typically be higher-wage occupations.

According to McKinsey forecasts, broader earning challenges lie ahead from a broader societal perspective. Although some demands for lower wage occupations will increase, a wide range of middle-income occupations will suffer the largest employment declines. As a result, income polarization may continue to expand.[339]

Economic sociologist Adam Hayes optimistically posits that while some industries and work roles will indeed fall as casualties of new technologies, they will be replaced by even greater, more open-ended opportunities:

- The automobile destroyed the horse and equestrian transportation industry. As the buggy makers and horse trainers saw their jobs disappear, many more new jobs were created in car factories, road and bridge construction and other industries.

- In the 19th century, when textile workers lost their jobs to mechanized looms, there were riots by the so-called Luddites who feared that the future was grim for labor.

- Elevator operators, once ubiquitous, were displaced by the automatic elevators we use today. In the 2000s, photographic film manufacturers were displaced by digital cameras.

- Eastman Kodak, which once employed many tens of thousands of workers, filed for bankruptcy and no longer exists.[340]

Computers are rapidly becoming smarter and are competing with practical roles we have naturally assumed needed us. Few occupations, including a major percent of highly-trained legal and medical diagnostic services, are immune from competitive AI and concomitant robotic automation workplace challenges.

Hedge funds are using AI to beat the stock market, Google is using it to diagnose heart disease more quickly and accurately and American Express is deploying bots to serve its customers online.

A 2016 McKinsey Global Institute study estimated that between 10 and 50 percent of all U.S. job tasks could be automated using existing robotic technology. In about 60 percent of 800 occupations surveyed, at least 30 percent of those primary activities can be replaced by software. Some jobs, such as driving and working in retail and fast-food, may become entirely obsolete.[341]

The McKinsey & Company's 2017 report concludes that whereas they had previously found that about half of the activities people are paid to do globally could theoretically be automated using currently demonstrated technologies, very few occupations—less than 5 percent—consisted of activities that can be fully automated.

However, in about 60 percent of occupations, at least one-third of the constituent activities could be automated. Their estimate implies that there will be substantial workplace transformations and changes for all workers.[342]

The McKinsey & Company researchers concluded that automation will have a lesser near-term effect on jobs that involve managing people, applying expertise and social interactions, where machines are currently less able to match human performance.

Automation will also have less impact on jobs in unpredictable environments—occupations such as gardeners, plumbers or providers of child and eldercare. Not only are they technically more difficult to automate, but they are economically less attractive from a business perspective because they typically command lower wages.

Among the most difficult activities to automate with current technologies are those that involve managing and developing people or that apply expertise to decision making, planning or creative work.[343]

Quantum Leaps Forward

Whereas the computational power of the human brain is largely pre-wired by evolution, AI capacities arising from revolutionary computer technology power and applications are growing exponentially.

Computer processing capacities are accelerating at an astounding rate with the advent of recent Quantum Computing (QC) advancements. As University of Maryland researcher Christopher Moore testified at an October 2017 House Science Committee hearing on American Leadership in Quantum Technology, merely 300 atoms under full quantum computer control might potentially store more pieces of information than the number of atoms that exist in the entire Universe.

Such implausible features are made possible by equally incomprehensible subatomic-scale phenomena. Unlike current computers which process tiny bits of data in a linear sequence as either a one or a zero, at the seemingly weird subatomic scale, a quantum bit (or qubit) can be both a zero and a one at the same time. As a result, rather than growing linearly, adding more qubits expands computing power exponentially.

QC progress now continues to rapidly accelerate following a three-decade scientific siesta since the concept was first proposed by Russian mathematician Yuri Manin in 1980:

- D-Wave Systems, a company based in Burnaby, British Columbia, demonstrated a special-function 16-qubit QC in 2007 at the Mountain View, California, Computer History Museum.

- In 2011, D-Wave Systems sold its first 128-qubit commercial system (D-Wave One) to the Lockheed Martin Quantum Computing Center located at the University of Southern California. The companies have since entered into multi-year agreements which have led to the development of more powerful D-Wave Two and D-Wave 2X systems.

- In 2013, Google established a Quantum Artificial Intelligence Laboratory (QAIL) at NASA's Ames Research Center at Moffett Field, California, in collaboration with the Universities Space Research Association (USRA).

- In 2015, QAIL publicly displayed a 10-foot-tall D-Wave 2X unit chilled at 180 times colder than deep space which is expected to operate 100 million times faster than any conventional computer.

- IBM has recently announced an initiative to build a commercially-available "IBM Q" along with an Application Program Interface (API) to enable customers and programmers to begin building interfaces between the company's existing five-qubit cloud-based computer and conventional computers.

Quantum systems are potentially capable of computational feats that have proven to be inconceivable with conventional technologies. For example, they might be used to model molecules which are ridiculously hard to model with a classical computer...trying to simulate the behavior of the electrons in even a relatively simple molecule which is enormously complex. IBM researchers used a quantum computer with seven qubits to model a small molecule made of three atoms.[344]

So, might artificial intelligence eventually evolve to duplicate or surpass the "real" intelligence that goes on in our brains?

Apart from our claims to unique spiritual, sensate and social qualities, can our human species keep up with—or perhaps even survive—exponentially growing artificial thinking machines and automated surrogates which are already outsmarting and outworking us in many areas of human endeavor?

Or on the other hand, might we not only learn from those products of our own innovation, but even interface AI systems with human minds to expand cognition and consciousness? Writing in the Wall Street Journal, Christof Koch believes that we can.

He urges:

> *There is one way to deal with this growing threat to our way of life. Instead of limiting further research into AI, we should turn it into an exciting new direction. To keep up with the machines we're creating, we must move quickly to upgrade our own organic computing machines: We must create technologies to enhance the processing and learning capabilities of the human brain.*[345]

Koch offers some early, yet promising neurotechnology examples that apply to present-day computational systems:

- Transcranial direct current stimulation: This noninvasive brain technology induces a weak electric field in the cortex underlying the skull. Research in animals and humans suggests that this may enhance neuro-plasticity, the process in which the brain improves its performance when an action is repeated over and over. Users wear headsets that gently stimulate the motor cortex while performing simple activities such as lifting weights, swinging a golf club or playing a piano.

- Electroencephalogram (EEG): Electrodes built into a headset detect brain waves during deep sleep. The device then plays low sounds that enhance the depth and strength of those waves, leading to more restful sleep. This noninvasive technology is presently limited because those billions of tiny nerve cells that generate brain waves are quite remote from the scalp, allowing only faint echoes of neuronal chatter to be picked up. Christof Koch concludes: "We aren't anywhere close to selectively silencing or amplifying the activity of small cliques of neurons."

- Neurosurgical Implants: Ultimately, to boost brain power we need to directly listen to and control individual neurons and atoms of perception, action, memory and consciousness. This currently requires some neurosurgery to penetrate the skull and access brain tissue. The good news is that brain-machine implant interfaces are happening faster than expected.

Nancy Smith was injured in a car accident which left her as a tetraplegic who can only move her shoulder and head. Neurosurgeons and neuroscientists implanted a tiny bed of nails in the region of her cortex to encode her intention to grasp a cup or press piano keys. Algorithms decode her neural signals and pass instruction to a musical synthesizer so that she can play music in her mind.

Bill Kochevar was paralyzed below the shoulders following a bicycle accident. A Cleveland-based team of doctors and neuroscientists placed electrodes into his left motor cortex that read out electrical tremors of about 100 neurons. From these, they decoded and transmitted his intentions to reach out and grasp objects by electronically stimulating muscles in his arm. While crude, it enables Kochevar to eat and drink by himself.

There are more than 50 patients with such neuronal listening devices installed in their brains.

Current and future applications include direct brain stimulation for obsessive-compulsive disorder, treatment-resistant depression, essential tremor, Parkinson's disease, epilepsy, stroke recovery and even blindness.

Christof Koch visualized that new neurotechnology developments may help patients recover lost functionality, including driving a car with their minds, plus a great deal more.

He contemplates:

> *My hope is that someday, a person could visualize a concept—say, the US Constitution. An implant in his visual cortex would read this image, wirelessly access the relevant online Wikipedia page and then write its content back into the visual cortex, so that he can read the webpage with his mind's eye. All of this would happen at the speed of thought.*[346]

Koch continues:

> *Another implant could translate a vague thought into a precise and error-free piece of digital code, turning anyone into a programmer. People could set their brains to keep focus on a task for hours on end, or control the length or depth of their sleep at will.*

Koch's vision brings a literal new meaning to the notions of "getting our heads together on an idea" and "sharing thoughts."

As he imagines this, he suggests:

> *Another exciting prospect is melding two or more brains into a single conscious mind by direct neuron-to-neuron links—similar to the corpus callosum, the bundle of two hundred million fibers that link the two cortical hemispheres of a person's brain. This entity could call upon the memories and skills of its members' brains but would act as one 'group' consciousness, with a single, integrated purpose to coordinate highly complex activities across many bodies.*

Christof Koch believes that humankind is at the threshold of a transformational new era that merges unlimited capacities of thinking machines and biological minds to revolutionize the entire meaning of intelligence:

> While the 20th century was the century of physics—think of the atomic bomb, the laser, the transistor—the 21st will be the century of the brain—the most complex piece of highly excitable matter in the known Universe.

Ultimately, our intelligent machines may even influence us to reinvent ourselves and equip us with "bigger brains" that will be needed to keep pace with our inventions.

As Harvard psychology professor Steven Pinker, the author of *The Stuff of Thought* reminds us in a New York Times opinion piece, new forms of media have always caused moral panics:

> The printing press, newspapers and television were all once denounced as threats to their consumers' brainpower and moral fiber. So too with electronic technologies. PowerPoint, we're told, is reducing discourse to bullet points. Search engines lower our intelligence, encouraging us to skim on the surface of knowledge rather than dive to its depths. Twitter is shrinking our attention spans.[347]

Pinker points out that while research shows that just as real-life learning experiences can "change the brain," the same applies to experiences gained through various media channels.

> Yes, every time we learn a fact or skill the wiring of our brain changes; it's not as if the

information is stored in the pancreas. But the existence of neural plasticity does not mean the brain is a blob of clay pounded into shape by experience.

He adds:

Media critics write that the brain takes on the qualities of whatever it consumes, the informational equivalent of "you are what you eat." As with primitive peoples who believe that eating fierce animals will make them fierce, they assume that watching quick cuts in rock videos turns your mental life into quick cuts or that reading bullet points and Twitter postings turns your thoughts into bullet points and Twitter postings.

Will Computers Control Us?

As elaborated in my recent book, *Reinventing Ourselves: How Technology is Rapidly and Radically Transforming Humanity,* we are at the earliest beginnings of a revolutionary new information era of incomprehensible societal scope and consequence.[348]

Many of these influences offer every appearance of being enormously constructive and positive.

The recent emergence of AI combined with the rapid growth of the Internet has already revolutionized the ways we seek, access, transfer and apply information available from a topically unlimited and geographically unbounded global library. These phenomena have had and will continue to have incalculably important and transformative influences upon most fields of human endeavor—those associated with academic online programs in particular.[349]

The Internet has made classroom walls and school buildings transparent, with technology essentially bringing the outside world in and inside learning out. It affords opportunities for people of all ages to earn college credits and degrees ranging from associate undergraduate to post-doctorate level from their homes to avoid the cost and inconvenience of moving or commuting to a physical campus community. Online students can arrange study times around parenting and work schedules.

Social media connections through the Internet have forever changed our senses of personal space, our life and work opportunities and countless aspects of our daily living routines. They have altered the ways we interact with loved ones, friends, employers and clients; have nurtured and supported a proactive business startup culture enabled through electronic commerce; and have opened up ways to identify and actively participate with groups and individuals who share our special interests and problems.

Since online shopping is removing the middleman, we can now purchase many of those products at much cheaper prices. Geographic distance is no longer a limit. The Internet has brought the power of online shopping and auctioning to people all over the world.

The Internet has freed us from geographic fetters and has connected us together in topic-based communities as a networked, globalized society connected by new technologies. It provides much of our news and pundit views. It connects us with real-time updates about happenings around the world, including real-time sports and weather information.

We also have new means to broadly communicate our own views and creativity through blogs and books published through e-publishers. We can read them online or on Kindle. We advertise and transact our goods and services across oceans and deserts.

We now have just-in-time information about almost

anything from anywhere. Print newspapers and magazines with dated information are being rapidly replaced by news-breaking electronic versions. Paperless record-keeping and publishing have replaced telephone books and cookbooks. Although not nearly as reliable, *Wikipedia* has replaced expensive and space-consuming sets of the *Encyclopedia Britannica*, which went out of print in 2011.

And with the advent of smartphones, we also no longer need wrist watches.

As blog writer Taryn Dentzel noted in an Openmind.com article, *How the Internet Has changed Everyday Life*:

> *The future of social communications will be shaped by an "always-online" culture. Always online is already here and will set the trend going forward. Total connectivity, the Internet, you can take with you wherever you go, is growing unstoppably. There is no turning back from global digitization.*[350]

Dentzel adds:

> *The Internet has turned our existence upside down. It has revolutionized communications, to the extent that it is now our preferred medium for everyday communication. In almost everything we do, we use Internet. Ordering a pizza, buying a television, sharing a moment with a friend, sending a picture over instant messaging. Before the Internet, if you wanted to keep up with the news, you had to walk down to the newsstand when it opened in the morning and buy a local edition reporting what had happened the previous day. But today a click or*

two is enough to read your local paper and any news source from anywhere in the world, updated to the minute.

The rise of the Internet has sparked much discussion and debate regarding how online communication affects social relationships.

Taryn Dentzel refers to the impacts of social media as having created a new communication democracy where the real value is that you stay in touch from moment to moment with people who matter most to you:

> *Social media let you share experiences and information, they get people and ideas in touch instantly, without frontiers. Camaraderie, friendship, and solidarity—social phenomena that have been around us as long as humanity itself—have been freed from the conventional restrictions of space and time.*[351]

At the same time, the Internet is also connecting us, along with private information, to uninvited audiences.

Global populations, Americans included, are trading away more and more of their personal privacy for promises of increased convenience and security. Spy cameras are sprouting up on lampposts and rooftops everywhere, facial recognition systems can track each of our individual movements and Internet-connected "smart cities" are wiring private home appliances within municipal energy monitoring and eventual control networks.

A smart city goal is premised upon supporting better decisions about design, policy and technology on information from an extensive network of sensors that gather data on everything from air quality to noise levels to people's activities. Some plans call for all vehicles to be autonomous and shared.

Robots will then roam underground doing menial chores like delivering mail.

Many of the same large American tech companies that will provide these devices are working with China to establish widespread public monitoring and social media censorship programs there.

In any case, there is no way to turn back the clock of progress where even Einstein's space-time continuum takes on a new dimension of meaning. Unlike the speed of light, there are no known theoretical limits to computational intelligence.

As Christof Koch, chief scientist and president of the Allen Institute of Brain Science in Seattle, predicts, sweeping societal and economic influences of AI signal a fourth industrial revolution. He observes:

> *The first, powered by the steam engine, moved us from agriculture to urban societies. The second, powered by electricity, ushered in mass production and created consumer culture. The third, centered on computers and the Internet, shifted the economy from manufacturing into services.*[352]

Koch points out that before modern farm equipment and tractors came along, it took 30 times more people to farm one hundred acres than it does today. This has resulted in producing more food for growing populations at affordable prices.

And while the Model T turned out the lights on many professions including blacksmiths and carriage makers, its introduction of affordable automobiles through mass production created huge new demands for labor created by the steel, glass, rubber, textile trade and oil and gas industries.

Dystopian visions of massive AI-driven job losses are premature. Throughout history, employment adapted as machines gradually replaced more and more aspects of labor over time. While once again, this fourth revolution will eliminate some jobs, it will also create opportunities for new ones that will require and enable more people to think smarter.

There is no turning back the clock or holding back the advances on the myriad of ways that information technology—AI and Internet connectivity in particular—are changing not only our lifestyles, but our fundamental perceptions regarding the types of lifestyles we deem most desirable as well.

Enthusiastic proponents promise tantalizingly optimistic visions: daily new conveniences previously conceivable in the fertile imaginations of a fiction writer but decades, or even a few years or days, ago; personal living efficiencies and household economies that save precious time and money; enhanced mobility through shared on-demand transportation services that banish most private automobiles to rusty scrap heaps of oblivion; and safety from predatory behaviors of others through ubiquitous, ever-watchful interconnected security devices.

If desired, those same technologies afford opportunities for many of us to live and work pretty much wherever and whenever we prefer through digital telecommuting that connects us to clients and employers in an inherently space-less world of increasingly faceless virtual relationships. In all cases, we remain connected and dependent upon ever watchful and tirelessly obedient digital assistants, who, in turn, monitor, record and market private information we generously, often unwittingly, share.

So, on balance, how can each of us assess the ultimate cost-benefit tradeoffs between conveniences and economies and the inevitable encroachments upon privacy and independence? In what ways are these accelerating developments transforming personal, business and societal cultures? Should we be compliant and adaptive, and even positively hopeful—or rather...well actually, there really is no constructive "rather."

Just as with every other major evolutionary game changer, and this is clearly a formative and consequential one, let's get used to the idea.

Maybe it will turn out just fine after all.

Larry Bell

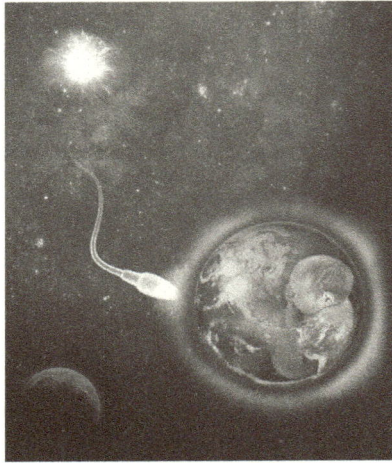

Endnotes

[1] *Out of Chaos, Evolution from the Big Bang to Human Intellect,* Wayne M. Bundy, Universal Publishers, 2007.

[2] *A Short history of Nearly Everything,* Bill Bryson, New York: Broadway Books, 2003.

[3] *The Selfish Gene,* Richard Dawkins, Oxford University Press, 1989.

[4] *Physics and Philosophy,* Werner Heisenberg, Harper Torchbooks, 1958.

[5] *The Evolution of Physics,* Albert Einstein and Leopold Infeld, Simon and Shuster, 1961.

[6] *S-Matrix Interpretation of Quantum Theory,* Henry Stapp, Lawrence Berkeley Laboratory, June 22, 1970.

[7] *Physics and Philosophy,* Werner Heisenberg, Harper Torchbooks, 1958.

[8] *Across the Frontiers,* Werner Heisenberg, Harper & Row, 1974.

[9] *Philosophiae Naturalis Principia Mathematica,* Isaac Newton, Sir Isaac Newton's Mathematical Principles of Natural Philosophy and His System of the World, 1946.

[10] Proceedings of the Royal Society of London, Correspondence of R. Bentley, Gary Zukav, HarperCollins, 2001.

[11] *Physics and Philosophy,* Werner Heisenberg, Harper Torchbooks, 1958.

[12] *S-Matrix Interpretation of Quantum Theory,* Henry Stapp, Lawrence Berkeley Laboratory, June 22, 1970.

[13] *The Dancing Wu Li Masters: An Overview of the New Physics*, Gary Zukav, HarperCollins, 2001.

[14] *S-Matrix Interpretation of Quantum Theory,* Henry Stapp, Lawrence Berkeley Laboratory, June 22, 1970.

[15] *The Evolution of Physics,* Albert Einstein and Leopold Infeld, Simon and Shuster, 1961.

[16] *Warped Passages, Unraveling the Mysteries of the Universe's Hidden Dimensions*, Lisa Randall, New York: ECCO, an imprint of Harper Collins, 2005

[17] *The Big Questions: Probing the promise and limits of Science*, Richard Morris, New York: Henry Holt and Company, 2002.

[18] *Origins of Existence: How Life Emerged in the Universe*, Fred Adams, New York: The Free Press, 2002.

[19] *The Big Questions: Probing the promise and limits of Science*, Richard Morris, New York: Henry Holt and Company, 2002.

[20] Ibid.

[21] *The Dancing Wu Li Masters: An Overview of the New Physics*, Gary Zukav, HarperCollins, 2001.

[22] *S-Matrix Interpretation of Quantum Theory*, Henry Stapp, Lawrence Berkeley Laboratory, June 22, 1970.

[23] *The Dancing Wu Li Masters: An Overview of the New Physics*, Gary Zukav, HarperCollins, 2001.

[24] Ibid.

[25] *Mind, Matter and Quantum Mechanics: The Copenhagen Interpretation*, Henry Pierce Stapp, SpringerLink, 2018.

[26] *The Dancing Wu Li Masters: An Overview of the New Physics*, Gary Zukav, HarperCollins, 2001.

[27] *The Whole Shebang: A State-of-the Universe(s) Report*, Timothy Ferris, New York: A Touchstone Book, 1998.

[28] *The Big Questions: Probing the promise and limits of Science*, Richard Morris, New York: Henry Holt and Company, 2002.

[29] Ibid.

[30] *Misconceptions about the Big Bang,* Charles Lineweaver and Tamara Davis, Scientific American, March 2005.

[31] *Origins of Existence: How Life Emerged in the Universe,* Fred Adams, New York: The Free Press, 2002.

[32] *Unseen Universe,* Corey Powell and Jessica Antola, A Discover Special, Winter, 2007.

[33] *The Big Questions: Probing the promise and limits of Science,* Richard Morris, New York: Henry Holt and Company, 2002.

[34] *Origins of Existence: How Life Emerged in the Universe,* Fred Adams, New York: The Free Press, 2002.

[35] *Out of Chaos, Evolution from the Big Bang to Human Intellect,* Wayne M. Bundy, 2007, Universal Publishers.

[36] *The Whole Shebang: A State-of-the Universe(s) Report,* Timothy Ferris, New York: A Touchstone Book, 1998.

[37] *Our growing, breathing galaxy,* Bart Wakker and Phillipp Richter, Scientific American, 2003.

[38] *On the difficulties of making Earth-like planets,* Stuart Ross Taylor, The Leonard Award Address, Meteorites & Planetary Science, V.34, pp.317-29, 1999.

[39] *Much ado about nothing, unseen Universe,* Michael S. Turner, a Discovery Special, Winter 2006.

[40] *Refugees for life in a hostile Universe,* Guillermo Gonzalez, Donald Brownie, and Peter D. Ward, Scientific American, 2001.

[41] *Extinction: Bad Gene or Bad Luck?,* David M. Raup, New York: W.W. Norton & Company, 1992.

[42] *Wonderful Life: The Burgess Shale and the Natural History,* Stephen J. Gould, New York: W.W. Norton 7 Company, 1989.

[43] *A Short History of Nearly Everything,* Bill Bryson, New York: Broadway Books, 2003.

[44] *Mystery of Mysteries: Is Evolution a social Construction?,* Michael Ruse, Cambridge, Massachusetts: Harvard University Press, 1999.

[45] Ibid.

[46] *Shadows of Forgotten Ancestors: A Search for Who We Are,* Carl Sagan and Ann Druyan, New York: Ballantine Books, 1992.

[47] *The Selfish Gene,* Richard Dawkins, Oxford; Oxford University Press, 2003.

[48] *Up from Dragons: The Evolution of Human Intelligence,* John R. Skoyles and Dorion Sagan, New York: McGraw Hill, 2002.

[49] *On the difficulties of making Earth-like planets,* Stuart Ross Taylor, The Leonard Award Address, Meteorites & Planetary Science, V.34, pp.317-29, 1999.

[50] *Out of Chaos, Evolution from the Big Bang to Human Intellect,* Wayne M. Bundy, 2007, Universal Publishers.

[51] *Why Darwin Matters: the Case Against Intelligent Design,* Michael Shermer, New York: Times books, Henry Holt and Company, 2006.

[52] *Mapping Human History: Discovering the Past Through Our Genes,* Steve Olson, Boston, Houghton Mifflin Company, 2002.

[53] *Shadows of Forgotten Ancestors: A Search for Who We Are,* Carl Sagan and Ann Druyan, New York: Ballantine Books, 1992.

[54] *What is Life?,* Robert Hazen, New Scientist, November 19-24, 2006.

[55] *Challenging Nature: The Clash of Science and Spirituality at the New Frontiers of Life,* Lee M. Silver, New York: HarperCollins Publishers, 2006.

[56] *Intimate Strangers: Unseen Life on Earth,* Cynthia Needham, Mahlon Hoagland, Kenneth McPherson, and Bert Dodson, Washington, D.C.: ASM Press, 2000.

[57] *Life as We Do Not Know It,* Peter D. Ward, New York: Viking, 2005.

[58] *The Brain,* Richard Restak, New York: Bantam Books, 1984.

[59] *What is Life,* Lynn Margulis and Dorion Sagan, Berkely: University of California Press, 2000.

[60] *The Cooperative Gene: How Mendel's Demon Explains the Evolution of Complex Beings,* Mark Ridley, New York: The Free Press, 2001

[61] *The River of Consciousness,* Oliver Sacks, New York-Toronto: Alfred A. Knopf, 2017.

[62] Ibid.

[63] Ibid.

[64] Ibid.

[65] Ibid..

[66] *The Ancestor's Tale: A Pilgrimage to the Dawn of Evolution,* Richard Dawkins, Boston: Houghton Mifflin Company, 2004.

[67] *Omphalos: An Attempt to Untie the Geological Knot*, Philip Henry Gosse, London, John van Voorst, 1857.

[68] *The River of Consciousness,* Oliver Sacks, New York-Toronto: Alfred A. Knopf, 2017.

[69] *The Double Helix,* James D. Watson, New York: Atheneum, 1968.

[70] *The Ages of Gaia: A Biography of Our Living Earth,* James Lovelock, edited by Lewis Thomas, New York: W.W. Norton & Company, 1995.

[71] *Out of Chaos, Evolution from the Big Bang to Human Intellect,* Wayne M. Bundy, 2007, Universal Publishers.

[72] Ibid.

[73] Ibid.

[74] *Life and Death of Planet Earth,* Peter Ward and Donald Brownlee, Rare Earth, 2003.

[75] Ibid.

[76] *Out of Chaos, Evolution from the Big Bang to Human Intellect,* Wayne M. Bundy, 2007, Universal Publishers.

[77] *The River of Consciousness,* Oliver Sacks, New York-Toronto: Alfred A. Knopf, 2017.

[78] Ibid.

[79] Ibid.

[80] Ibid.

[81] Ibid.

[82] Ibid.

[83] Ibid.

[84] *The Big Questions: Probing the Promise and Limits of Science,* Richard Morris, New York: Henry Holt and Company, 2002.

[85] *The Origin of Consciousness in the Breakdown of the Bicameral Mind,* Julian Jaynes, Boston-New York: First Mariner Books Edition, Houghton Mifflin Company, 2000.

[86] *Darwinism, an Exposition of the Theory of Natural Selection,* (Wallace's Contributions to the Theory of Natural Selection), Ch. 10), London: Macmillan, 1889.

[87] *The Brain: The Story of You,* David Eagleman, New York, Vintage Books, A division of Random House, LLC, 2015.

[88] *Brain Story: Unlocking our Inner World of Emotions, Memories, and Desires,* Susan Greenfield, New York: Dorling Kindersley publishing,

Inc., 2001.

[89] *Incognito: The Secret Lives of the Brain,* David Eagleman, New York: Vintage Books, A division of Random House, LLC, 2011.

[90] *The Brain: The Story of You,* David Eagleman, New York, Vintage Books, A division of Random House, LLC, 2015.

[91] Ibid.

[92] Ibid.

[93] *Incognito: The Secret Lives of Brains,* David Eagleman, New York: Vintage Books, a division of Random House, Inc., 2011.

[94] *Wider Than the Sky: The Phenomenal Gift of Consciousness,* Gerald M. Edelman, New Haven, London, Yale University Press, 2005.

[95] *Incognito: The Secret Lives of Brains,* David Eagleman, New York: Vintage Books, a division of Random House, Inc., 2011.

[96] *Constraints on cortical and thalamic projections: The no-strong-loops hypothesis*, F.H. Crick and C. Koch, Nature 391 (6664), 1998.

[97] *The Brain: The Story of You,* David Eagleman, New York, Vintage Books, A division of Random House, LLC, 2015.

[98] Ibid.

[99] *Thinking Whole: Rejecting Half-witted Left & Right Brain Limitations*, Larry Bell, Stairway Press, 2018.

[100] Ibid.

[101] *The Brain: The Story of You,* David Eagleman, New York, Vintage Books, A division of Random House, LLC, 2015.

[102] *Cosmic Musings: Contemplating Life Beyond Self,* Larry Bell, Stairway Press, 2016.

[103] *A Theory of Therapy, Personality Relationship as Developed in the Client-Centered Framework,* Carl Rogers. In (Ed.) S Koch, Psychology: A Study of Science, Vol. 3: Formulations of the person and the social context, New York: McGraw Hill, 1959.

[104] *The Human Advantage,* Suzana Herculano-Houzel, Cambridge, London: The MIT Press, 2016.

[105] *The Dragons of Eden: Speculations on the Evolution of Human Intelligence,* Carl Sagan, New York: Random House Publishing Group, Ballantine Books, 1977.

[106] *The Human Advantage,* Suzana Herculano-Houzel, Cambridge, London: The MIT Press, 2016.

[107] *Out of Chaos, Evolution from the Big Bang to Human Intellect*, Wayne M. Bundy, 2007, Universal Publishers.

[108] *An Ancestor to Call Our Own,* Kate Wong, Scientific American, June, 2003.

[109] *The Dragons of Eden: Speculations on the Evolution of Human Intelligence,* Carl Sagan, New York: Random House Publishing Group, Ballantine Books, 1977.

[110] *Shadows of Forgotten Ancestors: A Search for Who We Are,* Carl Sagan and Ann Druyan, New York: Ballantine Books, 1992.

[111] *Out of Chaos, Evolution from the Big Bang to Human Intellect*, Wayne M. Bundy, 2007, Universal Publishers.

[112] *Civilization Left its Mark on our Genes,* Bob Holmes, New Scientist, December 24-January 6, 2006.

[113] *Sapiens: A Brief History of Humankind,* Yuval Noah Harari, New York-London-Toronto-Sydney-New Delhi-Auckland: Harper Perennial, 2015.

[114] *The Dragons of Eden: Speculations on the Evolution of Human Intelligence,* Carl Sagan, New York: Random House Publishing Group, Ballantine Books, 1977.

[115] *Sapiens: A Brief History of Humankind*, Yuval Noah Harari, New York-London-Toronto-Sydney-New Delhi-Auckland: Harper Perennial, 2015.

[116] *Civilization Left its Mark on our Genes,* Bob Holmes, New Scientist, December 24-January 6, 2006.

[117] *Sapiens: A Brief History of Humankind,* Yuval Noah Harari, New York-London-Toronto-Sydney-New Delhi-Auckland: Harper Perennial, 2015.

[118] Ibid.

[119] *The role of the prefrontal Cortex in Dynamic Filtering*, Ralph Adolphs, Psychology, 2000.

[120] *The Evolution of Imagination,* Stephen T. Asma, University of Chicago Press, 2017.

[121] Ibid.

[122] *Neocortex Size as a Constant on Group Size in Primates*, Robin Dunbar, Journal of Human Revolution, 1992.

[123] *Understanding Primate Brain Evolution,* R.I.M Dunbar and Susanne

Shultz, Philosophical Transactions of the Royal Society of London B: Biological Sciences, 2007.

[124] *Our Brains they are A-Changing,* Mason Inman, New Scientist, September 17-23. 2005.

[125] *Emotional Intelligence: Why it Can Matter More than IQ,* Daniel Goleman, New York: Bantam Books, 2006.

[126] *Shadows of Forgotten Ancestors: A Search for Who We Are,* Carl Sagan and Ann Druyan, New York: Ballantine Books, 1992.

[127] The Age of the Spiritual Machines, Ray Kurzweil, Viking, 1999.

[128] *Sapiens: A Brief History of Humankind,* Yuval Noah Harari, New York, London, Toronto, Sydney, New Delhi, Auckland, Harper Perennial, 2015.

[129] Ibid.

[130] *Wisdom of the West,* Bertrand Russell, London: Bloomsbury Books, 1989.

[131] *The Evolution of Imagination: An Archaeological Perspective,* Steven J. Mithen, SubStance 30, 2001.

[132] *Shells of the French Aurignacian and Perigordian,* Yvette Taborin, in Before Lascaux: The Complete Record of the Early Upper Paleolithic, ed. Heidi Knecht, Anne Pike-Tay and Randall White, Boca Raton: CRC Press, 1993.

[133] *Sapiens: A Brief History of Humankind,* Yuval Noah Harari, New York, London, Toronto, Sydney, New Delhi, Auckland, Harper Perennial, 2015.

[134] *Double Child Burial from Sunghir (Russia): Pathology and Inferences for Upper Paleolithic Funerary Practices,* Vincenzo Formicola and Alexandra P. Buzhilova, American Journal of Physical Anthropology, 2004.

[135] *Sapiens: A Brief History of Humankind,* Yuval Noah Harari, New York-London-Toronto-Sydney-New Delhi-Auckland: Harper Perennial, 2015.

[136] *The Prehistory of the Mind: A Search for the Origins of Art,* Religion and Science, Steven Mithen, London: Thames and Hudson Ltd., 1996.

[137] *The Evolution of Imagination,* Stephen T. Asma, University of Chicago Press, 2017.

[138] Ibid.

[139] *The Age of Insight: The Quest to Understand the Unconscious in Art, Mind and Brain: From Vienna to the Present*, Eric Kandel, Random House, 2012.

[140] *The Evolution of Imagination,* Stephen T. Asma, University of Chicago Press, 2017.

[141] *An Early Bone Tool Industry from the Middle Stone Age at Blombos Cave, South Africa; Implications for the Origins of Modern Human Behavior*, Symbolism and Language, C.S. Henshilwood, F. d'Errico, C.W. Marean, R.G. Milo, and R. Yates, Journal of Human Evolution, 2001.

[142] *Sapiens: A Brief History of Humankind,* Yuval Noah Harari, New York, London, Toronto, Sydney, New Delhi, Auckland, Harper Perennial, 2015.

[143] *The Evolution of Imagination: An Archaeological Perspective,* Steven J. Mithen, SubStance 30, 2001.

[144] *The Origins of Greek Thought,* Jean-Pierre Vernant, New York: The Penguin Press, 1984.

[145] *Behavioral and Neural Correlates of Delay of Gratification 40 Years Later*, B.J. Casey et al., Proceedings of the National Academy of Sciences, 2011.

[146] *The Evolution of Imagination,* Stephen T. Asma, University of Chicago Press, 2017.

[147] *When Did Man Discover Fire?: Ancestors of Modern Humans Used Fire 350,000 Years Ago, New Study Suggests*, International Business Times, December 15, 2014.

[148] *Sapiens: A Brief History of Humankind,* Yuval Noah Harari, New York, London, Toronto, Sydney, New Delhi, Auckland, Harper Perennial, 2015.

[149] *Out of Chaos: Evolution from the Big Bang to Human Intellect,* Wayne Bundy, Boca Raton: Universal Publishers, 2007.

[150] *The Road to Now: Taking Stock of Evolution and Our Place in the World*, Melvin Bolton, Allen & Unwin: Crows Nest NSW, Australia, 2001.

[151] *Sapiens: A Brief History of Humankind,* Yuval Noah Harari, New York, London, Toronto, Sydney, New Delhi, Auckland, Harper

Perennial, 2015.

[152] *Out of Chaos: Evolution from the Big Bang to Human Intellect,* Wayne Bundy, Boca Raton: Universal Publishers, 2007.

[153] *The Substance of Civilization: Materials and Human History from the Stone Age to the Age of Silicon,* Stephen Sass, New York: Arcade Publishing, 1998.

[154] *Before the Dawn: Recovering the Lost History of Our Ancestors,* Nicholas Wade, New York: The Penguin Press, 2006.

[155] *Sapiens: A Brief History of Humankind,* Yuval Noah Harari, New York, London, Toronto, Sydney, New Delhi, Auckland, Harper Perennial, 2015.

[156] *The Prehistory of the Mind,* Steven Mithen, London: Thames and Hudson, 1999.

[157] *Out of Chaos: Evolution from the Big Bang to Human Intellect,* Wayne Bundy, Boca Raton: Universal Publishers, 2007.

[158] *Wisdom of the West,* Bertrand Russell, London: Bloomsbury Books, 1989.

[159] *The Substance of Civilization: Materials and Human History from the Stone Age to the Age of Silicon,* Stephen Sass, New York: Arcade Publishing, 1998.

[160] *The Artful Universe: The cosmic Source of Human Creativity,* John D. Barrow, Boston: Back Bay Books, Little Brown, and Company, 1998.

[161] *The World Economy, Vol. 2,* Angus Maddison, Paris: Development Centre of Organization of Economic Cooperation and Development, 2006.

[162] *Sapiens: A Brief History of Humankind,* Yuval Noah Harari, 2015, HarperCollins, Harper Perennial.

[163] Ibid.

[164] Ibid.

[165] *A History of Knowledge: Past, Present, and Future,* Charles Van Doren, New York: Carol Publishing Company, 1991.

[166] *The Lost Civilizations of the Stone Age,* Richard Rudgley, New York: Simon & Schuster, 2000.

[167] *Sapiens: A Brief History of Humankind,* Yuval Noah Harari, 2015, HarperCollins, Harper Perennial.

[168] Ibid.

[169] *Martyrdom and Persecution in the Early Church,* W H Frend, Cambridge: James Clarke & Co., 2008.

[170] *Sapiens: A Brief History of Humankind,* Yuval Noah Harari, 2015, HarperCollins, Harper Perennial.

[171] *Why I am a Hindu,* Shashi Tharoor, New Delhi: Aleph Book Company, 2018; as summarized in the Wall Street Journal, *How Hinduism Has Persisted for 4,000 Years,* January 18, 2019.

[172] *Axialism and Empire,* Sheldon Pollock, in Axial Civilizations and World History, ed. Johann P. Arnason, S.N. Eisenstadt and Bjorn Wittock, Leiden: Brill, 2005.

[173] *China: A History,* Harold M. Tanner, Indianapolis: Hackett Publishing Company, 2009.

[174] *A Short List of Moral Values,* Richard T. Kinnier er al., Counseling and Values, October 2000, Vol.45.

[175] *Guns, Germs and Steel: The Fates of Human Societies,* Jared Diamond, New York: W.W. Norton and Company, 1999.

[176] *The Genius of China: 3,000 Years of Science, Discovery, and Invention*, Robert Temple, New York: Simon and Schuster, 1986.

[177] *Doubt and Certainty,* Tony Rothman and George Sudarshan, Reading: Perseus Books, 1998.

[178] *Connections,* James Burke, Boston: Little, Brown and Company, 1995.

[179] *The Passion of the Western Mind: Understanding the Ideas that Have Shaped Our World View*, Richard Tarnas, New York: Ballantine Books, 1991.

[180] Ibid.

[181] *Wisdom of the West,* Bertrand Russell, London: Bloomsbury Books, 1989.

[182] Ibid.

[183] *The Origins of Greek Thought,* Jean-Pierre Vernant, Ithaca: Cornell University Press, 1984.

[184] *Wisdom of the West,* Bertrand Russell, London: Bloomsbury Books, 1989.

[185] *The Origins of Greek Thought,* Jean-Pierre Vernant, Ithaca: Cornell University Press, 1984.

[186] *Wisdom of the West,* Bertrand Russell, London: Bloomsbury Books, 1989.

[187] Ibid.

[188] *Cleopatra & Antony,* Brian Haughton, Ancient History Encyclopedia (article), January 10, 2011.

[189] Ibid.

[190] *The Passion of the Western Mind: Understanding the Ideas that Have Shaped Our World View*, Richard Tarnas, New York: Ballantine Books, 1991.

[191] *The Crusades: Motivations, Administration, and Cultural Influence,* Rachel Rooney with Andrew Miller, Digital Collections for the Classroom, The Newberry.org, October10, 2016.

[192] *The History of the Rise and Fall of the Roman Empire,* Edward Gibbon, (Six volumes), London: Strahan & Cadell, 1776-1789.

[193] *History of the Byzantine Empire,* Alexander Vasiliev, University of Wisconsin Press, 1952.

[194] *Wisdom of the West,* Bertrand Russell, London: Bloomsbury Books, 1989.

[195] *Landmarks in Western Science: From Prehistory to the Atomic Age*, Peter Whitfield, New York: Rutledge, 1999.

[196] *The Passion of the Western Mind: Understanding the Ideas that Have Shaped Our World View*, Richard Tarnas, New York: Ballantine Books, 1991.

[197] *What the History Books Left Out,* Ehsan Masood, New Scientist, April 1-7, 2006.

[198] *The Passion of the Western Mind: Understanding the Ideas that Have Shaped Our World View*, Richard Tarnas, New York: Ballantine Books, 1991.

[199] *Landmarks in Western Science: From Prehistory to the Atomic Age*, Peter Whitfield, New York: Rutledge, 1999.

[200] *The Passion of the Western Mind: Understanding the Ideas that Have Shaped Our World View*, Richard Tarnas, New York: Ballantine Books, 1991.

[201] *The Birth of Modern Science,* Paolo Rossi, 2001, Blackwell Publishing. ISBN 0631227113.

[202] *Leonardo da Vinci: Italian Artist, Engineer and Scientist,*

Encyclopedia Britannica.
https://www.britannica.com/biography/Leonardo-da-Vinci.
[203] *Leonardo da Vinci, Artist, Inventor and Universal Genius of the Renaissance,* http://www.leonardo-history.com/life.htm?Section=S6; http://www.history.com/topics/leonardo-da-vinci.
[204] *The Life and Times of Leonardo,* Liana Bartolon, 1967, Paul Aamlyn, London, ISBN 075251587X.
[205] *The Passion of the Western Mind: Understanding the Ideas that Have Shaped Our World,* Richard Tarnas, New York: Ballantine Books, 1991.
[206] *Landmarks in Western Science: From Prehistory to the Atomic Age,* Peter Whitfield, New York: Rutledge, 1999.
[207] *The Search for the Beginning and End of the Universe,* John Seife, New York: Penguin Books, 2003.
[208] *The Whole Shebang: A State-of-the-Universe(s) Report,* Timothy Ferris, New York: A Touchstone Book, 1998.
[209] *The Passion of the Western Mind: Understanding the Ideas that Have Shaped Our World,* Richard Tarnas, New York: Ballantine Books, 1991.
[210] Ibid.
[211] *Robert Hooke,* Encyclopedia.com, Complete Dictionary of Scientific Biography, Charles Scribner's Sons, 2008.
[212] *Landmarks in Western Science: From Prehistory to the Atomic Age,* Peter Whitfield, New York: Rutledge, 1999.
[213] *A Brain for all Seasons: Human Evolution & Abrupt Climate Change,* William H. Calvin, Chicago: The University of Chicago Press, 2002.
[214] *Wisdom of the West,* Bertrand Russell, London: Bloomsbury Books, 1989.
[215] *What is Enlightenment?,* Immanuel Kant, *In The Portable Enlightenment Reader.* Edited by Isaac Krannick, New York, Penguin Books, 1995.
[216] *The Whole Shebang: A State-of-the-Universe(s) Report,* Timothy Ferris, New York: A Touchstone Book, 1998.
[217] *Thinking Whole,* Larry Bell, Stairway Press, 2018.
[218] *The Enlightenment,* Josh Rahn, The Literature Workshop,

http:www.online-literature.com, 2011.

[219] *The Unity of Knowledge,* Edward O. Wilson, New York: Alfred A. Knopf, 1998.

[220] *Wisdom of the West,* Bertrand Russell, London: Bloomsbury Books, 1989.

[221] *Out of Chaos: Evolution from the Big Bang to Human Intellect,* Wayne Bundy, Boca Raton: Universal Publishers, 2007.

[222] *The Unbound Prometheus,* David S. Landes, Cambridge, Massachusetts: Press Syndicate of the University of Cambridge, 1969, ISBN 978-0-521-09418-4.

[223] *English and American Tool Builders,* Joseph Wickham Roe, New Haven, Connecticut, Yale University, 1916, Reprinted by Mcgraw-Hill, New York and London, 1926 and by Lindsay Publications, Bradley, Illinois.

[224] *The Wealth and Poverty of Nations,* David Landes, 1999, New York: W.W. Norton and Company.

[225] *Empire of Cotton: A Global History,* Sven Beckert, 2014, New York: U.S. Vintage Books Division of Penguin Random House.

[226] *English and American Tool Builders,* Joseph Wickham Roe, New Haven, Connecticut, Yale University, 1916, Reprinted by Mcgraw-Hill, New York and London, 1926 and by Lindsay Publications, Bradley, Illinois.

[227] *Inventing the Cotton Gin: Machine and Myth in Antebellum America*, Angela Lakwete, 2005, Baltimore: John Hopkins University Press.

[228] *The Unbound Prometheus,* David S. Landes, Cambridge, Massachusetts: Press Syndicate of the University of Cambridge, 1969, ISBN 978-0-521-09418-4.

[229] *Technological Transformations and Long Waves,* Robert Ayres, 1989, http://pure.iiasa.ac.at/id/eprint/3225/1/RR-89-001.pdf

[230] *A History of Industrial Power in the United States,* Louis Hunter, 1985, Vol. 2: Steam Power, Charlottesville: University Press of Virginia.

[231] *A History of Metallurgy, Second Edition,* R.F. Tylecote, 1992: London: Maney Publishing, for the Institute of Materials.

[232] *The Unbound Prometheus,* David S. Landes, Cambridge,

Massachusetts: Press Syndicate of the University of Cambridge, 1969, ISBN 978-0-521-09418-4.

[233] *The Most Powerful Idea in the World: A Story of Steam Industry and Invention*, William Rosen, 2012, Chicago: University of Chicago Press.

[234] *The Unbound Prometheus*, David S. Landes, Cambridge, Massachusetts: Press Syndicate of the University of Cambridge, 1969, ISBN 978-0-521-09418-4.

[235] *From the American System to Mass Production, 1800-1932: The Development of Manufacturing Technology in the United States*, David A. Hounshell, Baltimore: Johns Hopkins University Press.

[236] *Lectures on Economic Growth*, Robert Lucas, 2002, Cambridge: Harvard University Press.

[237] *Technological Transformations and Long Waves*, Robert Ayres, 1989, http://pure.iiasa.ac.at/id/eprint/3225/1/RR-89-001.pdf

[238] *Sapiens: A Brief History of Humankind*, Yuval Noah Harari, New York, London, Toronto, Sydney, New Delhi, Auckland: HarperCollins/Harper Perennial, 2015.

[239] *Hyperspace: A Scientific Odyssey through Parallel Universes, Time Warps, and the 10th Dimension*, Michio Kaku, 1994, New York: Anchor Books, Doubleday.

[240] *An Encyclopedia of the History of Technology*, Ian McNeil, 1990, London: Routledge. ISBN 0-415-14792-1.

[241] *The Education of Thomas Edison*, Jim Powell, February 1, 1995, Foundation for Economic Education.

[242] Ibid.

[243] Ibid.

[244] Ibid..

[245] *Thinking Whole: Rejecting Half-Witted Left & Right Brain Limitations*, Larry Bell, 2018, Arizona: Stairway Press ISBN 978-1-949267-02-0.

[246] *S-Matrix Interpretation of Quantum Theory*, Henry Stapp, June 22, 1970, Lawrence Berkeley Laboratory.

[247] *The Evolution of Physics*, Albert Einstein and Leopold Infeld, 1961, Simon and Shuster.

[248] *Thinking Whole: Rejecting Half-Witted Left & Right Brain*

Limitations, Larry Bell, 2018, Arizona: Stairway Press ISBN 978-1-949267-02-0.

[249] *The Dancing Wu Li Masters: An Overview of the New Physics*, Gary Zukav, 2001, HarperCollins.

[250] *Taking Flight: Inventing the Aerial Age, from Antiquity through the First World War*, Richard P. Hallion, 2003, New York: Oxford University Press ISBN 0195160355.

[251] *Flops of famous inventions,* George L. Dowd, Popular Science, 1930.

[252] *Helicopter Development in the Early Twentieth Century,* Judy Rumerman, Centennial Flight Commission. https://www.centennialofflight.net/essay/Rotary/early_20th_century/HE2.htm.

[253] *Kaman K-225,* R.D. Connor and R.E. Lee, July 21, 2001, Smithsonian National Air and Space Museum.

[254] *Eisenhower on the Opportunity Cost of Defense Spending,* November 12, 2007, as reported by Scott Horton in Harper's Magazine, March 2, 1019.

[255] *The Major Alliances of World I Began as Hope for Mutual Protection,* Robert Wilde, ThoughtCo, https://www.thoughtco.com/world-war-one-the-major-alliances-1222059.

[256] *The Major Alliances of World I Began as Hope for Mutual Protection,* Robert Wilde, ThoughtCo, https://www.thoughtco.com/world-war-one-the-major-alliances-1222059.

[257] *Germany and the Causes of the First World War*, Mark Hewitson, 2004, London: Bloomsbury.

[258] *When and Why Did Germany Enter World War I?,* Alexander Bauer, August 17, 2018, Military History and Wars of Germany, Quora.com.

[259] *All Quiet on the Western Front,* Erich Maria Remarque, 1929, Ballantine Books.

[260] *Crumbling of Empires and Emerging States: Czechoslovakia and Yugoslavia as (Multi)national Countries*, International Encyclopedia of the First World War (WWI), https://encyclopedia.1914-1918-

online.net/article/crumbling_of_empires_and_emerging_states_czec
hoslovakia_and_yugoslavia_as_multinational_countries

[261] *Nazis and Slavs: From Racial Theory to Racist Practice,* John
Connelly, 1999, Cambridge: Central European University,
JSTOR 4546842.

[262] *The Impact of Nazism: New Perspectives on the Third Reich and Its
Legacy,* Alan E Steinweis and Daniel E. Rogers, 2003, Lincoln &
London: University of Nebraska Press, ISBN 9780803242999.

[263] *The Munich Agreement—September 1938,* Richard Nelson,
September 21, 2018, The Guardian.

[264] *The Road to World War II: How Appeasement Failed to Stop
Hitler,* Klaus Wiegrefe, February 9, 2009, Spiegel Online.

[265] *BBC—History—World Wars: Invasion of Poland,* Bradley
Lightbody, March 30, 2011, UK: BBC.

[266] *Why Finland allied itself with Nazi Germany | Letter,* Geoffrey
Roberts, February 23, 2018, UK: The Guardian.

[267] *Fort Eben Emael,* Tim Saunders, January 1, 2005, Casemate
Publishers.

[268] *Swedish neutrality during the Second World War: tactical success
or moral compromise?,* Paul A. Levine, 2002, In Wylie,
Neville. European Neutrals and Non-Belligerents During the Second
World War, Cambridge: Cambridge University Press.

[269] BBC on this Day, 15, 1940: Victory for RAF in Battle of Britain,
September 15, 2005, UK: BBC.

[270] *Death in the Snow: Battle of Moscow,* Kennedy Hickman,
September 19, 2018, ThoughtCo.

[271] *Early Warnings: How American Journalists Reported the Rise of
Hitler,* Jennie Rothenberg Gritz, March 3, 2012, The Atlantic.

[272] *Pendulum of War: The Three Battles of El Alamein,* Niall Barr,
2004, Overlook Press. ISBN 9781585676552.

[273] *The Making of Modern Japan,* Marius Jansen, 2002, Cambridge
Mass: Harvard University Press.

[274] *Wuhan, 1938: War, Refugees, and the Making of Modern China,*
Stephen MacKinnon, 2008, University of California Press. ISBN 978-
0520254459.

[275] *Kaiigun: Strategy, tactics and technology in the Imperial Japanese*

Navy, 1887-1941, David C. Evans and Mark R. Peattie, 1997, Annapolis Maryland: Naval Institute Press.

[276] *Creating Military Power; the Sources of Military Effectiveness,* Risa Brooks and Elizabeth A. Stanley, 2007, Stanford University Press, ISBN 0-8047-5399-7.

[277] The Pacific War Online Encyclopedia.

[278] *Indian Independence Day: Everything You Need to Know about Partition between India and Pakistan 70 Years On,* Barney Henderson, August 15, 2017, the Telegraph.

[279] *How One Newspaper Report Changed World History,* Mark Dummett, December 16, 2011, BBC News.

[280] *African Political Leadership: Jomo Kenyatta, Kwame Nkrumah and Julius Nyerere, A.B. Assensoh,* 1998, Krieger Publishing Company.

[281] *Stalin's Wars From World War to Cold War, 1939-1953,"* Geoffrey Roberts, 2006, Cambridge, Mass: Yale University Press, ISBN 9780300112047.

[282] *The Moldavians: Romania, Russia, and the Politics of Culture,* Charles King, 2000, Hoover Institution Press, ISBN 9780817997922.

[283] *Red Century: From the 'October Revolution' in 1917, Communism swept the Globe,* Will Englund, October 26, 2016, The Washington Post.

[284] *Nuclear Forces Reduced While Modernizations Continue, Says SIPRI,* June 16, 2014, Stockholm International Peace Research Institute.

[285] *Skunks, Bogies, Silent Hounds, and the Flying Fish: The Gulf of Tonkin Mystery,* Robert J. Hanyok, August 2-4, 1964, the Wayback Machine, Cryptologic Quarterly, Winter2000/Spring 2001 Edition, Vol. 19, No. 4/Vol. 20, No. 1.

[286] *Gulf of Tonkin: McNamara admits it didn't happen,* March 4, 2008, You Tube.

[287] *Norodom Sihanouk, Cambodian Leader Through Shifting Alliances, Dies at 89,* Elizabeth Becker and Seth Mydans, October 14, 2012, The New York Times.

[288] *Forty years on from the fall of Saigon: Witnessing the end of the Vietnam War,* Martin Wollacott, April 21, 2015, The Guardian.

[289] http://biography.com/people/wernher-von-braun-9224912.

[290] Chertok interview in Izvestia, Nos. 54, 55, 56, 57, 58, March 4-9, 1992.

[291] Ibid.

[292] Ibid.

[293] Ibid.

[294] *The Rocket Team,* Frederick Ordway and Mitchell R. Sharpe, 1979, New York: Thomas Y. Crowell.

[295] Ivanovsky interview with James Hartford on Jan. 26, 1993 in his book *Korolev; How One Man Masterminded the Soviet Drive to Beat America to the Moon,* Wiley & Sons, Inc., 1997, p. 80.

[296] *The Practical Significance of Konstantin Tsilokovsky's Proposals in the Field of Rocketry,* Sergei P. Korolev, 1986 paper presented at a meeting commemorating Tsiolkovsky's 100th birthday, Sept. 17, 1957, USSR Academy of Sciences, Moscow, reprinted in English in History of the USSR; New Research, 5, Social Sciences Today, Moscow.

[297] *Creative Legacy of Academician Sergi Pavlovich Korolev,* M.V. Keldysh, editor, and G.S. Vetrov, compiler,(in Russian), Moscow, Nauka, 1980.

[298] Ibid.

[299] Ibid.

[300] Ibid.

[301] Ibid.

[302] *Recollections of Childhood: Early Experiences in Rocketry as Told by Wernher von Braun,* 1963, MSFC History Office, NASA Marshall Space Flight Center. http://biography.com/people/wernher-von-braun-9224912.

[303] Ibid.

[304] Ibid.

[305] Ibid.

[306] Ibid.

[307] Ibid.

[308] Ibid.

[309] Ibid.

[310] Ibid.

[311] *Tass*, September 8, 1958.

[312] *Corona: America's First Spy Satellite Program,* Quest, Wayne Day,

Grand Rapids, MI: Cspace Press, Summer, 1995.

[313] *US Reconnaissance Satellite Programs,* Jonathan McDowell and Robert A. McDonald, *Corona Success for Space Reconnaissance,* PE & RS Photogrammetric Engineering and Remote Sensing, June 1995.

[314] *Pravda,* November 29, 1977.

[315] *Tass,* September 8, 1958.

[316] National Intelligence Estimate, The Russian Space Program, Dec.5, 1962, as reported in Hartford, James, *Korolev: How One Man Masterminded the Soviet Drive to Beat America to the Moon,* Wiley & Sons, Inc., 1997.

[317] *The Practical Significance of Konstantin Tsilokovsky's Proposals in the Field of Rocketry,* Sergi P. Korolev, paper presented at a meeting commemorating Tsiolkovsky's 100th birthday, Sept. 17, 1957, USSR Academy of Sciences, Moscow, reprinted in English in History of the USSR; New Research, 5, Social Sciences Today, Moscow, 1986.

[318] *The Way it Was: the Difficult Fate of the N-1 Project,* M. Rebrov, Krasnaya Zvezda (Russian), Jan. 15, 1990, No.11, p.4, as reported in Hartford, James, *Korolev: How One Man Masterminded the Soviet Drive to Beat America to the Moon,* Wiley & Sons, Inc., 1997.

[319] *Creative Legacy,* Kelydysh and Vetrov, 1997, as reported in Hartford, James, Korolev: How One Man Masterminded the Soviet Drive to Beat America to the Moon, Wiley & Sons, Inc.

[320] *SP-4209 The Partnership: A History of the Apollo-Soyuz Test Project,* NASA Space History Office, https://history.nasa.gov/SP-4209/ch2-4.htm.

[321] *There are not enough resources to support the world's population,* John Guillebaud, June 10, 2014, https://www.abc.net.au/radionational/programs/ockhamsrazor/there-are-not-enough-resources-to-support-the-worlds-population/5511900.

[322] *Should Artificial Intelligence Copy the Brain?,* Christopher Mims, Wall Street Journal, August 4-5, 2018.

[323] Ibid.

[324] *Learning Representations for Multimodal Data with Deep Belief Nets,* Nitish Srivastava and Ruslan Salakhutdinov, Presented at the ICML Representation Learning Workshop, Edinburgh, Scotland, UK

2012, http://www.cs.toronto.edu/~nitish/icml2012/paper.pdf.
[325] Moore's Law is based on calculations done by Carver Mead at Moore's request. "While pursuing these researches, Mead responded to a query from Intel-founder Gordon Moore about the possible size of microelectronic devices. Mead provided the empirical analysis behind Moore's law (predicting a doubling of computer power every 18 months)."
http://worrydream.com/refs/Mead%20-%20American%20Spectator%20Interview.html
[326] *How cloud computing is changing the world...without you knowing*, Joe Baguley, September 24, 2013, The Guardian.com, Media network blog.
[327] *Cloud Computing; Current and Future Impact on Organizations,* Yiyun Zhu, March 20, 2017, Western Oregon University, Department of Computer Science student thesis paper, Digital Commons@WOU.
[328] *Cloud Computing is Crucial To The Future Of Our Societies—Here's Why,* Joy Tan, Feb. 25, 2018, Forbes.com.
[329] *Harnessing automation for a future that works,* Lames Manyika, Michael Chui, Mehdi Miremadi, Jacques Bughin, Katy George, Paul Willmott, and Martin Dewhurst, McKinsey & Company/ McKinsey Global Institute, January, 2017.
[330] *Where machines could replace humans- and where they can't (yet),* Michael Chui, James Manyika, and Mehdi Miremadi, McKinsey Quarterly, July 2016.
[331] *The history of computing is both evolution and revolution,* Justin Zobel, The Conversation.com, May 31, 2016.
[332] Where do we humans still hold the advantage? According to Kevin McCaney who writes for *Governmentciomedia.com, "there are quite a few mental and physical activity areas."*
[333] *Where Do Humans Outperform AI?,* Kevin McCaney, Governmentciomedia.com, April 4, 2018.
[334] Ibid.
[335] *Will AI Best All Humans Tasks by 2060? Experts Say Not So Fast,* Alisa Valudes Whyte, Huffington Post, June 25, 2017.
[336] Ibid.
[337] *Industrial robots will replace manufacturing jobs—and that's a good*

thing, Matthew Randall, TechCrunch.com, October 9, 2016.

[338] *Jobs lost, jobs gained: What the future of work will mean for jobs, skills and wages,* McKinsey & Company, James Manyika, Susan Lund, Michael Chui, Jacques Bughin, Jonathan Woetzel, Paul Batra, Ryan Ko, and Saurabh Sanghvi, November, 2017.

[339] Ibid.

[340] *20 Industries Threatened by Tech Disruption*, Adam Hayes, Investopedia, Feb. 6, 2015.

[341] *Where machines could replace humans— and where they can't (yet)*, Michael Chui and James Manyika, 2016, McKinsey Quarterly.

[342] *Jobs lost, jobs gained: What the future of work will mean for jobs, skills and wages,* McKinsey & Company, James Manyika, Susan Lund, Michael Chui, Jacques Bughin, Jonathan Woetzel, Paul Batra, Ryan Ko, and Saurabh Sanghvi, November, 2017.

[343] *Where machines could replace humans— and where they can't (yet),* Michal Chui, James Manyika, and Mehdi Miremadi, McKinsey Quarterly, McKinsey & Company, July 20, 2016.

[344] *10 Breakthrough Technologies*, MIT Technology Review, 2018.

[345] *We'll Need Bigger Brains Keeping Up With the Machines*, Christof Koch, Wall Street Journal Review Section, October 28-29, 2017.

[346] Ibid.

[347] *Mind Over Mass Media,* Steven Pinker, June 10, 2010, The New York Times.

[348] *Reinventing Ourselves: How Technology is Rapidly and Radically Transforming Humanity*, Larry Bell, 2019, Stairway Press.

[349] *How the Internet Has changed Everyday Life*, Zaryn Dentzel, Openmind.com.

[350] Ibid.

[351] Ibid.

[352] https://alleninstitute.org/what-we-do/brain-science/about/team/staff-profiles/christof-koch/